Power Electronics and Power Systems

The Power Electronics and Power Systems book series encompasses power electronics, electric power restructuring, and holistic coverage of power systems. The series comprises advanced textbooks, state-of-the-art titles, research monographs, professional books, and reference works related to the areas of electric power transmission and distribution, energy markets and regulation, electronic devices, electric machines and drives, computational techniques, and power converters and inverters. The series features leading international scholars and researchers within authored books and edited compilations. All titles are peer reviewed prior to publication to ensure the highest quality content.

To inquire about contributing to the series, please contact:

Dr. Joe Chow
Administrative Dean of the College of Engineering and Professor of Electrical, Computer and Systems Engineering
Rensselaer Polytechnic Institute
Jonsson Engineering Center, Office 7012
110 8th Street
Troy, NY USA
Tel: 518-276-6374
chowj@rpi.edu

Pengwei Du

Renewable Energy Integration for Bulk Power Systems

ERCOT and the Texas Interconnection

 Springer

Pengwei Du
The Electric Reliability Council of Texas (ERCOT)
Taylor, TX, USA

ISSN 2196-3185 ISSN 2196-3193 (electronic)
Power Electronics and Power Systems
ISBN 978-3-031-28641-4 ISBN 978-3-031-28639-1 (eBook)
https://doi.org/10.1007/978-3-031-28639-1

This Springer imprint is published by the registered company Springer Nature Switzerland AG
The registered company address is: Gewerbestrasse 11, 6330 Cham, Switzerland

Preface

The Electric Reliability Council of Texas (ERCOT) is the Independent System Operator (ISO) that operates the electric grid and manages the deregulated wholesale electricity market for the ERCOT region. It manages the power flow to about 24 million of Texas customers, representing approximately 90% of the state electricity demand. As the independent system operator, ERCOT schedules power on an electric grid that connects more than 46,500 miles of transmission lines and over 710 generation units. ERCOT has about 77 GW of total generating capacity and a peak load of 74,820 MW recorded on August 12, 2019. There are more than 1800 active market participants that generate, buy, sell, or use wholesale electricity in ERCOT market. The total market size is around $38 billion based on 381 billion kWh market volume and average $0.10/kWh rate. ERCOT is connected to Eastern Interconnect (EI) through two high-voltage direct current (HVDC) ties rated at 220 MW and 600 MW. ERCOT is also connected to the Comision Federla de Electricidad (CFE), the electric grid in Mexico, through three HVDC ties rated at 36 MW, 100 MW, and 150 MW. This arrangement makes ERCOT an electrical "island" with only limited asynchronous interconnections. As a result of low natural gas prices and rapid development of wind and solar generation in the past 20 years, ERCOT's generation mix has dramatically changed from being heavily dominated by coal and gas-steam generation in the late 1990s to now having a substantial share of renewable generation and simple and combine cycle generation.

Texas also has rich wind and solar resources. To date, the total installed capacity for wind and solar resources is 35,644 MW and 11,440 MW, respectively. Most of wind and solar generation projects in ERCOT are interconnected at transmission level. As a result of the rapid development of wind and solar generation in west Texas and very scarce transmission capacity in the region, there was initially insufficient transmission capacity to transfer wind and solar energy to the large load centers in the north and central of the state. This situation could cause substantial wind energy curtailments when the penetration of wind generation was high. In 2005, the Texas legislature ordered the Public Utility Commission of Texas (PUCT) to determine areas where future renewables would likely to develop, so called "competitive renewable energy zones" (CREZ). The commission designated ERCOT to develop a transmission plan to interconnect these areas to the existing transmission grid and elevate already-existing transmission congestion from the West to the North and Central parts of Texas. As a result, about 3600 right-of-way miles of 345 kV transmission lines were constructed with a capability of accommodating 18.5 GW of wind generation capacity. The project was completed in December 2013 with a cost of approximately $6.8 billion.

As renewable resources in the ERCOT region are far from the load centers where most of conventional synchronous generation is also concentrated, adequate voltage support must be maintained, and associated voltage stability must be assessed to ensure reliable operation of the ERCOT grid. Wind and solar resources are connected to the grid through inverters. Inverter-base resources (IBRs) such as wind and solar, being low to zero marginal price resources, displace more expensive conventional synchronous generation in unit commitment and dispatch. However, these rotating turbine generators and motors which are synchronously connected to the system store kinetic energy and can release stored kinetic energy to slow down the frequency decay in response to a sudden loss of generation. This is called inertial response. Inertial response provides an important contribution to reliability in the initial moments following a generation or load trip event. However, non-synchronous resources like wind and solar generation do not contribute to the synchronous inertia. As the system inertia is declining, the grid could be susceptible to the reliability risk if there is lack of frequency responsive capacity at the time when the inertia is low.

ERCOT is operating an energy-only wholesale electricity market. In 1999, Senate Bill 7 (SB7) restructured the Texas electricity market by unbundling the investor-owned utilities and creating retail customer choice in those areas. In 2001, ERCOT began its single control area operation and opened both its wholesale and retail electricity market to competition based on a zonal market structure. On December 1, 2010, ERCOT successfully launched the locational marginal pricing-based nodal market. ERCOT also procures ancillary services to ensure extra capacity is available to address variability that cannot be covered by the five-minute real-time energy market. Ancillary services in ERCOT currently include responsive reserve service, regulation service, non-spinning reserve service, black start, and emergency inter-ruptible loads. The latter two services are used in emergency and system recovery conditions, while the former three services are used to balance net load, which is defined as system load minus wind and solar power production. As more renewable resources are connected to the ERCOT grid, the grid reliability need is evolving, which leads to the increase in the ancillary service quantities and the need to re-design ancillary service market to maintain sufficient system reserves and flexibility during real time grid operations.

In the electricity grid, at any moment balance must be maintained between electricity consumption and generation. However, renewable resources, in contrast to conventional generation systems, are not dispatchable. To optimally schedule other dispatchable resources, ERCOT needs to estimate how much power wind and solar generation resources will produce in the future. Renewable production potential (RPP) forecasting is essential to the integration of large amount of variable renewable generation. ERCOT acquired its first wind forecasting service in 2010 and procured multiple wind and solar forecasting services over the years to provide centralized forecasts for all renewable generation resources connected to ERCOT transmission grid.

As more wind and solar generation was built, ERCOT introduced a number of changes to the market rules, wholesale energy, and ancillary service markets and operation practices, and also developed new situational awareness tools and real-time studies to ensure continued system reliability. The operation experience at ERCOT has shed some light on what the technical challenges will be and how to overcome these challenges when the penetration of renewable resources becomes high. This also shows that a high penetration of renewable resources can be both technically feasible and economically viable for a future power grid.

This book provides a thorough understanding of the basic aspects that need to be addressed for a reliable operation at a high penetration of renewable resources as well as describes the most recent technologies and solutions developed at ERCOT to address these challenges. This thorough understanding can reveal the trend of the reliability issues that will grow over time and it allows the most cost-effective long-term solutions to be implemented before these issues become prominent. Effective system operation can benefit the market efficiency and improve the system's capability to integrate more renewable resources.

This book covers a variety of subjects associated with the interconnection of renewable resources and presents a number of innovative and practical solutions developed at ERCOT. This book also emphasizes the interrelation between the economic aspects, reliability, and regulatory requirements since any successful strategies helping to improve the security of a future grid with renewable resources present must be cost-effective and the enhancement of these strategies need to be supported by the regulatory agencies in charge of the grid security and reliability.

This book could be useful for engineers and operators in power system planning and operation, as well as academic researchers. It can serve as an excellent introduction for university students in electrical engineering at both undergraduate and postgraduate levels. The dissemination of the knowledge contained in this book can stimulate more ideas and innovations to be developed and eventually help to facilitate the interconnection of more renewable resources in the future grid.

This book is divided into ten chapters.

Chapter 1 provides an overview of ERCOT and recent development of renewable resources at ERCOT. This chapter covers a variety of operational challenges with high inverter-based resource penetration.

Chapter 2 describes in depth the fundamental principles of the wholesale market which ERCOT is operating and illustrates through examples how the market is operated to clear the price. A well-designed wholesale electricity market can produce appropriate pricing signals to reflect the scarcity conditions and create a long-term incentive to the development of new resources.

Chapter 3 discusses the customized market solutions implemented at ERCOT to improve the efficiency and reliability, i.e., to allow renewable resources to fully participate in the both day-ahead and real-time market operation.

Chapter 4 presents the overview of ancillary services and the methodology to determine the need of these ancillary services. Ancillary services are necessary to support the transmission of electric power from generators to consumers given the obligations of control areas and transmitting utilities within those control areas to maintain reliable operations of the interconnected transmission system.

Chapter 5 discusses a new primary frequency control market where both generators and load resources with under-frequency relays can participate. This will incentivize more resources which can contribute to the system reliability when needed and reward them based on their actual performance. In the long term, it will attract more participants to compete, and the market efficiency will be eventually improved.

Chapter 6 presents a new ancillary service framework being implemented at ERCOT in order to address the primary frequency control issues associated with the declining system inertia. This new design can balance the need for both reliability and economics while opening the ancillary service market to energy-limited resources like batteries.

Chapter 7: "System Inertia Trend and Critical Inertia" describes the efforts undertaken at ERCOT to track the trend of the historical inertia and to develop tools and methods to mitigate any negative impact of low inertia conditions that could arise in the future.

Chapter 8 introduces an iterative linear-based approach to solve the multiple-hour reactive scheduling problem, i.e., to determine when and how to switch the discrete reactive devices for bulk power systems, in the presence of a large amount of renewable resources.

Chapter 9 discusses the ERCOT's experiences of using the wind and solar forecasting system to maintain the grid reliability, and describes the specifics for the renewable forecast, data flow, system architecture, and forecast performance evaluation.

Chapter 10 presents an ensemble machine learning-based method to forecast wind power production, which uses both the wind generation forecasted by a numerical weather prediction (NWP) model and the meteorological observation data from weather stations.

The author is grateful to the help provided by the colleagues at ERCOT for the development of these innovative works. The future of renewable resources is promising as they offer many benefits to the grid and the society. The intent of this book is to encourage more people to contribute to this dynamic field and enable further exploitation of new revolutions for renewable integration.

Finally, the author appreciates the staff at Springer for their assistance and help in the preparation of this book.

Austin, TX, USA   Pengwei Du

Contents

1 Renewable Integration at ERCOT 1
  1 Overview 1
    1.1 Texas Power System 1
    1.2 Wind Generation Development in Texas 5
    1.3 Solar Generation Development in Texas 7
    1.4 Interconnection Requirements for Generation Resources 7
  2 Transmission Development and Capacity Adequacy 9
    2.1 Transmission Access 9
    2.2 Transmission Reinforcement 9
    2.3 Operational Challenges with High Inverter-Based
        Resource Penetration 11
    2.4 Capacity Adequacy and Wind Generation Resources 12
  3 ERCOT Energy and Ancillary Services Market 13
    3.1 Ancillary Services 15
  4 Reliability and Security of Grid Operations 17
    4.1 Requirement for Primary Frequency Response 18
    4.2 ERCOT Frequency Performance 18
    4.3 System Inertia Trend 19
    4.4 Renewable Generation Forecasting 20
    4.5 Operations Analysis and Studies 21
  5 Conclusions 25
  References 25

2 Overview of Market Operation at ERCOT 27
  1 Overview 27
    1.1 Network Modeling 28
    1.2 Day-Ahead Operations 30
    1.3 Adjustment Period Operations 30
    1.4 Real-Time Operations 31

    2   Day-Ahead Market (DAM)............................... 31
        2.1   DAM Overview..................................... 31
        2.2   ERCOT and QSE Activities in DAM.................. 34
        2.3   DAM Engine...................................... 43
    3   RUC... 47
        3.1   RUC Process Overview............................. 47
        3.2   Input to RUC Process.............................. 49
        3.3   RUC Process...................................... 53
        3.4   Energy Offer Curve for RUC-Committed Resource 63
    4   Real-Time SCED....................................... 63
        4.1   SCED: Energy Dispatch............................ 64
        4.2   Load Frequency Control (LFC)...................... 72
    5   Conclusions.. 73
    References... 73

3   Market Designs to Integrate Renewable Resources 75
    1   Overview... 75
    2   ERCOT Nodal Market.................................. 77
    3   Short-Term Wind/PVGR Generation Forecasting
        and Current Operating Plan (COP)....................... 78
    4   IRR Scheduling in DAM, RUC, and SCED................. 78
        4.1   IRR Scheduling in DAM............................ 78
        4.2   IRR Generation Scheduling in RUC................... 79
        4.3   IRR Generation Scheduling in Real-Time SCED 81
        4.4   Base Point Deviation Charge for IRR................. 84
        4.5   Effect on Management of Congestion.................. 85
        4.6   Effect on Market Prices............................ 86
    5   Management of Generic Transmission Constraints
        (GTCs)... 86
        5.1   Management of GTC Limits........................ 86
        5.2   Non-to-Exceedance (NTE) Method................... 89
        5.3   Implementation of NTE Concept..................... 91
    6   Incorporation of 5-Minute Wind/Solar Ramp
        into SCED... 94
        6.1   Generation to Be Dispatched (GTBD)................. 95
        6.2   Forecast of 5-Minute Solar Ramp.................... 96
    7   Conclusions.. 100

4   Ancillary Services (AS) at ERCOT 103
    1   Overview... 103
        1.1   Responsive Reserve Service (RRS)................... 103
        1.2   Regulation Reserve Service......................... 104
        1.3   Non-spinning Reserve Service....................... 104
    2   Regulation Services.................................... 104
        2.1   Short-Term Wind Generation Forecasting................ 106

2.2   Method to Determine Regulation Services
        Requirement...   108
  2.3   Procedures to Determine Regulation Service
        Requirement...   112
3   Responsive Reserve Requirement...........................   118
  3.1   Quantification of PFR and FFR Requirement
        for Inertias..   118
  3.2   RRS Requirement..   122
4   Non-spinning Reserve.....................................   124
  4.1   Background..   124
  4.2   Net Load Forecast Error (NLFE) Analysis...............   126
  4.3   Net Load Ramp-Up..   129
  4.4   Adjustment to Non-Spin Need by Considering Forced
        Outage..   130
  4.5   Procedures to Determine Non-Spin Need................   131
5   Summary..   133
References..   135

**5   Design of New Primary Frequency Control Market for Hosting
      Frequency Response Reserve Offers from Both Generators and
      Loads..   137**
1   Introduction of Frequency Control........................   137
2   Impact of Renewable Resource over Inertia
    and Primary Frequency Control............................   139
3   Co-optimization of Energy and FRR in Day-Ahead
    Market..   140
  3.1   Day-Ahead Market Co-optimization Model..............   140
  3.2   Solution of Day-Ahead Market Co-optimization..........   145
  3.3   Case Studies...   148
4   Stochastic Formulations of Co-optimization of Energy
    and FRR in Day-Ahead Market.............................   154
  4.1   Energy, PFR, and Inertia Scheduling Without
        Uncertainties...   156
  4.2   Energy, PFR, and Inertia Scheduling Under
        Uncertainties...   157
5   Conclusions...   171
References..   172

**6   New Ancillary Service Market for ERCOT: Fast Frequency
      Response (FFR)..   177**
1   Introduction..   177
2   Existing Ancillary Service Market at ERCOT...............   179
  2.1   Regulation Service..   180
  2.2   Responsive Reserve Service (RRS)........................   181
  2.3   Non-spin Reserve Service (NSRS)........................   181

3   Inertia Trend and Primary Frequency Control
    at ERCOT.. 181
    3.1   Inertia Trend at ERCOT............................. 182
    3.2   Overview of Frequency Control Coordination
          at ERCOT... 183
    3.3   RRS at ERCOT Before Re-design of AS Market.......... 184
4   New Ancillary Service Market.............................. 184
5   Fast Frequency Response (FFR)............................. 187
    5.1   Qualifications of FFR and Performance Evaluations.... 189
    5.2   Telemetry Data Requirement for Deployment
          and Recall of FFR................................. 189
6   Maximum Amount of FFR Allowed............................. 190
7   Benefits of FFR... 192
    7.1   Impact of FFR over Critical Inertia................. 192
    7.2   RRS Cost Saving with FFR Resources.................. 193
8   Conclusions... 195
References.. 196

7   **System Inertia Trend and Critical Inertia**................ 199
    1   Basics of Synchronous Inertia......................... 199
    2   Inertia at ERCOT...................................... 200
    3   Historical Synchronous Inertia Trends................. 203
    4   Determining Critical Inertia.......................... 205
    5   ERCOT Tools to Monitor and Forecast System Inertia.... 210
    6   Impact of Parameter Changes on Critical Inertia....... 212
        6.1   "Faster" Frequency Response..................... 213
        6.2   "Earlier" Frequency Response.................... 213
        6.3   Lower UFLS Trigger.............................. 214
        6.4   Reduction of Largest Possible Loss of Generation. 217
    7   International Review of Inertia-Related Challenges
        and Mitigation Measures............................... 218
    8   Summary of Potential Mitigation Measures to Lower
        Critical Inertia or Support Minimum Inertia Level..... 218
    9   Conclusions... 222
    References.. 222

8   **Multiple-Period Reactive Power Coordination for Renewable**
    Integration.. 223
    1   Introduction.. 224
    2   Literature Review..................................... 225
    3   Renewable Integration and New Transmission
        Operators... 227
    4   Reactive Power Coordination (RPC) Tool................ 228
        4.1   Architecture of RPC............................. 228
        4.2   Objective Function of RPC....................... 230

    4.3   Constraints of RPC 231
    4.4   Mathematic Formulations of RPC 231
    4.5   Solution Methodology 233
  5 Special Considerations 233
    5.1   Sensitivity of Reactive Power for a Regulating Bus 233
    5.2   Reactive Device's Temporal Constraint 234
    5.3   Handling Special Capacitor Banks 234
  6 Case Studies 235
    6.1   ERCOT Network Model 235
    6.2   Verification of Temporal Constraints 236
    6.3   Simulation Results 238
  7 Conclusions 240
  References 241

9 Renewable Forecast 243
  1 Introduction 243
  2 Wind Forecasting System 244
    2.1   Wind Forecasting System Overview 244
    2.2   Data Flow of Wind Forecasting System 246
    2.3   Input Data for Wind Forecasting System 247
    2.4   Design Approach of Wind Forecasting System 248
  3 Solar Forecasting System and Forecast Errors 250
    3.1   Solar Forecasting System 250
    3.2   Solar Forecast Error Analysis 254
  4 Summary and Conclusions 262
  References 262

10 Ensemble Machine Learning-Based Wind Forecasting
    to Combine NWP Output with Data from Weather Stations 263
  1 Introduction 263
  2 Numerical Weather Prediction 266
  3 West Texas Mesonet (WTM) 268
    3.1   Value of WTM Data: An Example 269
  4 Machine Learning-Based Ensemble Method 270
  5 Experimental Results 274
    5.1   Performance of Three Machine Learning Algorithms 275
    5.2   Performance of Base Algorithms for Large
          Wind Ramp 277
    5.3   Performance of Ensemble Method 277
    5.4   Robustness of Proposed Method 278
  6 Conclusions 279
  References 279

Index 283

Chapter 1
Renewable Integration at ERCOT

1 Overview

1.1 Texas Power System

The Electric Reliability Council of Texas (ERCOT) is the independent system operator (ISO) that operates the electric grid and manages the deregulated wholesale electricity market for the ERCOT region. ERCOT is one of the ten independent system operators (ISOs) and regional transmission organizations (RTOs) of the ISO/RTO Council (IRC) in North America. The ten ISOs and RTOs in North America serve two-thirds of electricity consumers in the United States as well as more than 50% of Canada's population.

ERCOT is the electricity grid and market operator for the majority of the state of Texas. It manages the power flow to about 24 million of Texas customers, representing approximately 90% of the state electricity demand. As the independent system operator, ERCOT schedules power on an electric grid that connects more than 46,500 miles of transmission lines and over 710+ generation units [1]. ERCOT has 77 GW of total generating capacity and a peak load of 74,820 MW recorded on August 12, 2019. There are more than 1800 active market participants that generate, move, buy, sell, or use wholesale electricity in the ERCOT market. The total market size is around $38 billion based on 381 billion kWh market volume and average $0.10/kWh rate.

The remaining part of the state of Texas is served from either Southwest Power Pool (SPP) or the Midcontinent Independent System Operator (MISO), which are parts of the Eastern Interconnection (EI) or Western Electricity Coordinating Council (WECC). ERCOT is connected to EI through two high-voltage direct current (HVDC) ties rated at 220 MW and 600 MW. ERCOT is also connected to the Comision Federal de Electricidad (CFE), the electric grid in Mexico, through three HVDC ties rated at 36 MW, 100 MW, and 150 MW. As shown in Fig. 1.1, this

© The Author(s), under exclusive license to Springer Nature Switzerland AG 2023
P. Du, *Renewable Energy Integration for Bulk Power Systems*, Power Electronics and Power Systems, https://doi.org/10.1007/978-3-031-28639-1_1

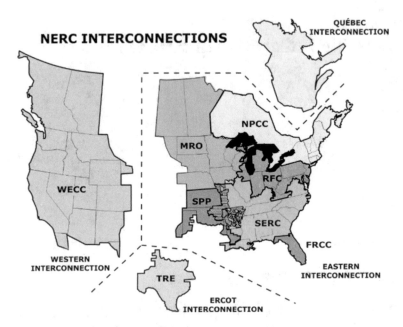

Fig. 1.1 NERC Interconnections map [2]

arrangement makes ERCOT an electrical "island" with only limited asynchronous interconnections.

As a result of low natural gas prices and the rapid development of wind generation, ERCOT's generation mix has dramatically changed in the past 20 years from being heavily dominated by coal and gas steam generation in the late 1990s to now having a substantial share of renewable generation and simple and combined cycle generation, as shown in Fig. 1.2. Furthermore, nearly 6700 MW of coal and gas steam units were retired in the past 4 years [3, 4].

The current generation mix consists of coal, nuclear, natural gas (steam generation, simple cycle and combined cycle combustion turbines), biomass, landfill gas, hydro, solar, and wind generation. The ERCOT system is heavily reliant on natural gas generation from both an installed capacity and energy perspective, the latter of which is shown in Fig. 1.3.

In parallel to the shift of the supply, it is also a challenge to support the highest peak during the peak day of the year. The peak load is due to coincidental usage and is driven by the end users. Figure 1.4a shows the impact of weather on load by customer type at ERCOT. On the left side, the ERCOT load at 5:00 p.m. on March 11, 2019, was 40,158 MW, and the temperature in Dallas was 63 °F. For such a load level, a large number of generators were shut down as idle. On the right side is a peak load of 74,898 MW at 5:00 p.m. on August 12, 2019, and the temperature in Dallas was 101 ° F. A similar comparison between the load compositions in the winter season is also shown in Fig. 1.4b. To meet this high load demand, both base load and peak load generation need to be online. In an unstructured utility, capital investments

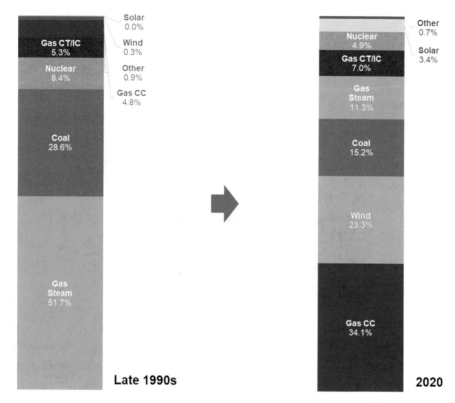

Fig. 1.2 Changes in ERCOT resource mix, in percent of installed capacity

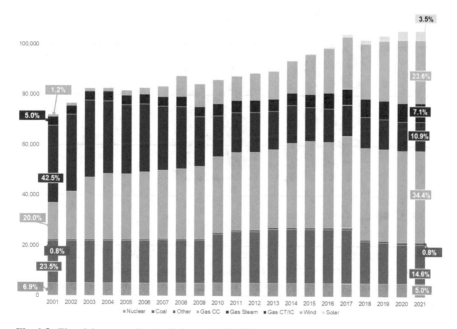

Fig. 1.3 Electricity generation by fuel type in ERCOT

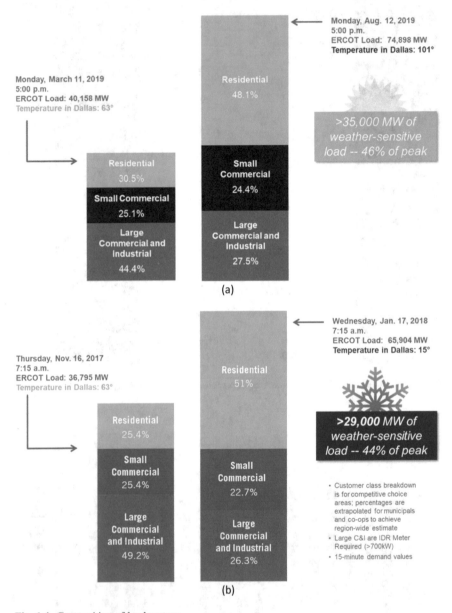

Fig. 1.4 Composition of load sectors

are driven up by the need to serve this peak load so that electrical infrastructure is underutilized for the majority of the time. In a power market environment, the decision of building new generation is a market behavior but is still largely influenced by the prospect of future load shape.

1.2 Wind Generation Development in Texas

One of the reasons for rapid wind generation development is that Texas has rich wind resources—see Fig. 1.5. The best wind potential is in the Texas Panhandle[1] and West Texas as well as along the Gulf Coast. These areas have an average wind generation capacity factor of 35–40%.

Apart from the availability of excellent wind resource, the renewable portfolio standard (RPS) and Production Tax Credit (PTC) greatly influence the development of wind generation in Texas [8]. An RPS is a mandate passed by the Texas Legislature to establish a minimum amount of renewable resources in the state's generation portfolio. Texas's RPS target was 10,000 MW of renewable generation by 2025. This target was exceeded by 2010. Additionally, the federal Production Tax Credit (PTC) has been vital in maintaining economic viability of the wind generation projects. The PTC is a federal incentive that provides financial support for the development of renewable energy facilities. It provides a 2.3 cent/kWh incentive for the first 10 years of a renewable facility operation. The PTC originally started in

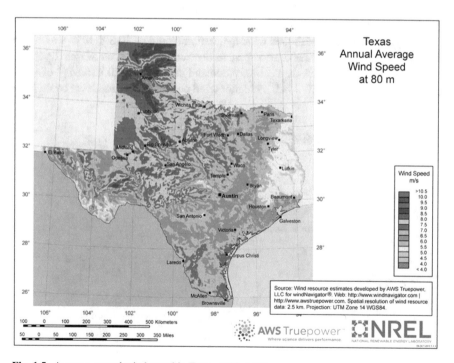

Fig. 1.5 Average annual wind speed in Texas at 80 m [5]

---

[1] The Texas Panhandle is a rectangular area of the state, consisting of the northernmost 26 counties in the state and bordered by New Mexico to the west and Oklahoma to the north and east.

1992 and expired at the end of 2013. Additionally, in December 2014, the PTC was retroactively extended for the projects that were under construction by the end of 2014. In order to receive tax credits, these projects had to be operational by January 1, 2017. These policy changes are clearly reflected in wind generation resources development in ERCOT, which has shown a steady increase of about 1000 MW per year from 2008 to 2014 and a substantial leap in 2015 (during which over 3 GW was added) followed by an additional increase of more than 5 GW planned by the end of 2016, as seen in Fig. 1.6.

At the end of December 2015, another extension of PTC was signed. This 5-year extension will run through 2019. This new retroactive PTC extension allows developers to earn the full 2.3 cent/kWh tax credit for projects that meet "commence construction" criteria in 2015 and 2016 and are completed within 2 years. The credit drops by 20% in 2017, 20% further in 2018, and a final 20% more in 2019 before terminating on January 1, 2020 [6].

The installed wind capacity within the ERCOT footprint is 24,971 MW (as of February 1, 2021), making Texas the leading state for wind capacity in the United States. Most of the wind generation resources in ERCOT are connected at transmission voltage levels (69 kV, 138 kV, or 345 kV). Project sizes range from 1 MW to 250 MW (110 MW on average).

The most recent instantaneous wind generation record of 22,893 MW was set on January 14, 2021. The instantaneous penetration record was set on March 22, 2021, with wind serving 66.47% of electricity demand at the time (20,986 MW of wind generation and 31,574 MW load). A higher wind power penetration level could have

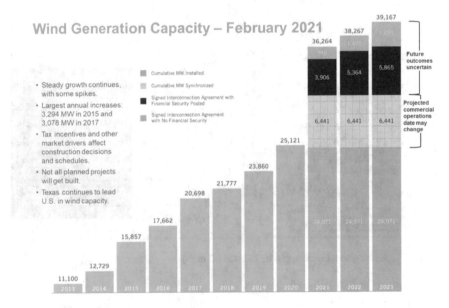

Fig. 1.6 ERCOT wind installations by year

been set at that time with wind generation potentially being sufficient to serve 78% of load; however, due to transmission constraints and thermal generators reaching their minimum sustainable generation levels, over 3635 MW of wind generation was curtailed. With rapid developments of wind generation projects, these records are likely to be exceeded in the near future.

About 89% of total installed wind generation capacity is located in the West Texas and Texas Panhandle. These areas are rich in wind resources (see Fig. 1.5), are scarcely populated, and thus are currently offering better economic opportunities for wind generation buildout. About 11% of capacity is along the Gulf Coast near and south of Corpus Christi. This area has more favorable wind patterns well correlated with ERCOT load and, at times, high wholesale electricity market prices compared to West Texas and the Texas Panhandle. However, the coastal area has a lot of industrial and tourist activity which makes wind generation development there more challenging and expensive.

1.3 Solar Generation Development in Texas

With intense sun and vast tracts of empty land that can accommodate the huge scale of major solar farms, West Texas has long been primed for rapid solar development. Texas's free market approach to electricity production and loose regulation of development encourages big electricity projects of any kind, including solar. With technological innovations, the cost of developing solar farms has dropped about 40% in Texas in the last 5 years, according to the Solar Energy Industries Association (SEIA). Once a solar farm is built, it's inexpensive to operate compared to gas and coal-fired plants, because its fuel is free. Federal tax credits have further cut the cost of developing solar farms. Meanwhile, demand for solar electricity has increased as both the public and corporations have embraced it as a means of battling the climate crisis. The state's utility-scale solar capacity increased to 5,777 megawatts by 150% in 2020. In 2021, installed solar capacity is expected to grow more than 130% to 13,449 megawatts—see Fig. 1.7. Even with solar's growth, it still makes up only about 2% of electricity generation in Texas. In comparison, natural gas makes up about 44%, wind about a quarter, and coal 16%.

1.4 Interconnection Requirements for Generation Resources

ERCOT ensures operational security through a strong set of interconnection requirements that apply to generating resources including wind and solar generation units. This set of requirements include but are not limited to:

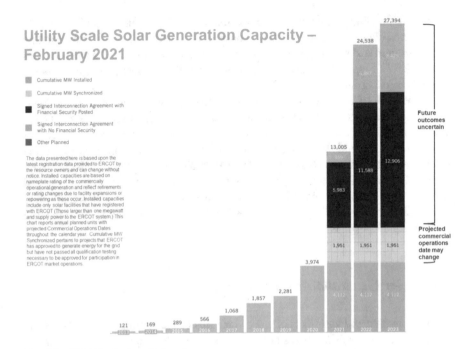

Fig. 1.7   ERCOT solar installations by year

- Voltage Support: To provide 0.95 power factor leading and lagging reactive capability determined at the maximum active power output and which has to be provided continuously.
- Frequency Support: To provide primary frequency response with maximum 5% droop and maximum frequency deadband of ±0.017 Hz when frequency responsive headroom is available (Fig. 1.8).
- Voltage and Frequency Ride Through Capability: To maintain the ability to remain connected to the grid and provide necessary support during voltage and frequency events of predefined severity.
- Dispatchability: To follow dispatch instruction to maintain and adjust output timely.
- Ramping: Each wind or solar resource shall limit its ramp rate to 20% per minute of its nameplate rating (MWs) as registered with ERCOT when responding to or released from an ERCOT deployment.
- Telemetry: All generation resources in ERCOT are required to provide numerous telemetry data points to ERCOT with a time resolution of 2–10 seconds. These data points are critical to operation studies and situational awareness tool as well as for offline analysis.

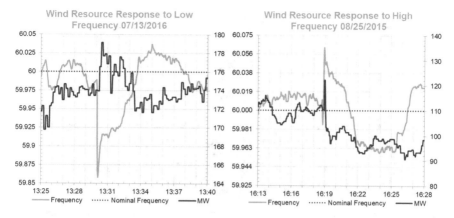

Fig. 1.8 Frequency response from wind generation resources in ERCOT

2 Transmission Development and Capacity Adequacy

2.1 Transmission Access

In ERCOT, generation owners are not required to pay for transmission upgrades necessary to facilitate the interconnection. The generation owner only pays for the connection from their facility to the nearest point of interconnection on the transmission grid. Any other transmission upgrades necessary to accommodate new generating capacity are paid by the demand customers based on a flat rate approved by the Public Utility Commission of Texas (PUCT). This reduces the cost of market entry for all new generation projects compared to some other systems where generators must pay for necessary transmission system upgrades associated with their project.

Figure 1.9 shows the process of approving and permitting new transmission lines or transmission upgrades. As electricity demand continues to increase in the region, ERCOT analyzes trends statewide to determine where new infrastructure is needed. Annual studies are conducted to determine system needs within a 6-year planning horizon. Longer-term studies are conducted every other year to ensure that near-term planning decisions are informed by long-term system trends.

2.2 Transmission Reinforcement

As a result of the rapid development of wind generation in West Texas and very scarce transmission capacity in the region, there was initially insufficient transmission capacity to transfer wind energy to the large load centers in the north and central of the state (i.e., Dallas, Austin, San Antonio, and Houston). This situation could cause substantial wind energy curtailments when the penetration of wind generation

Fig. 1.9 Process of approving and permitting new transmission lines or transmission upgrades

was high. In 2005, the Texas Legislature ordered the PUCT to determine areas where future renewables would likely to develop, so-called competitive renewable energy zones (CREZ) as shown in Fig. 1.10. The commission designated ERCOT to develop a transmission plan to interconnect these areas to the existing transmission grid and elevate already-existing transmission congestion from the west to the north and central parts of Texas. As a result, about 3600 right-of-way miles of 345 kV transmission lines were constructed with a capability of accommodating 18.5 GW of wind generation capacity. The project was completed in December 2013 with a cost of approximately $6.8 billion. The new lines are open-access, and their use is not limited to wind generation. Some of these transmission lines also benefit fast-growing demand from oil and gas industry in western Texas.

The CREZ transmission project also included new transmission facilities in the Texas Panhandle. Prior to the CREZ project, there were no ERCOT transmission lines extending into that area and therefore no load or generation in the area connected to ERCOT system. Furthermore, in the beginning of the CREZ project, there was no generation with signed interconnection agreements in the Panhandle area. The reactive equipment necessary to support the export of power from the Panhandle was implemented to accommodate 2400 MW of wind generation capacity, even though the transmission lines were constructed for much larger generation capacity. This decision was made because the size and location of any additional equipment would be dependent upon the size and location of the wind generation that eventually would be developed in the area in the future. By the time CREZ transmission project was completed, there was only 200 MW of wind generation capacity installed in the Panhandle. However, over the next 2 years (2014–2015), another 2.2 GW of wind generation capacity was built, reaching the limit that CREZ transmission facilities in the Panhandle area were designed for. There are currently an additional 5.3 GW of wind, 0.2 GW of solar, as well as 0.2 gas-fired generation capacity with signed interconnection agreements planned in the Panhandle area in the 2016–2017 timeframe [7].

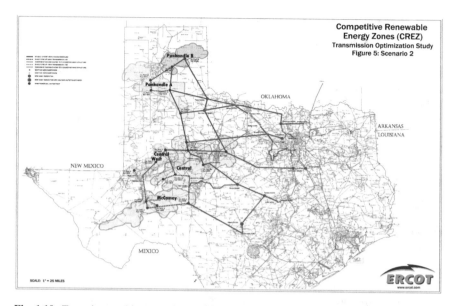

Fig. 1.10 Texas invested in transmission lines linking wind-rich "competitive renewable energy zones" to the state's largest cities [9]

The Panhandle part of the ERCOT grid is remote from synchronous generators, and wind generation projects in the Panhandle are equipped with advanced power electronic devices that further weaken the system strength due to limited short-circuit current contributions. As a consequence, voltage control becomes very difficult because a small reactive power change results in large voltage deviations. By the end of 2015, additional transmission reinforcement projects were proposed to provide reactive power support and allow further wind generation growth in the Panhandle area.

2.3 Operational Challenges with High Inverter-Based Resource Penetration

To date, the highest percentage of load served by wind generation in ERCOT is 59.3%, on May 2, 2020, 2:10 a.m., with a total of 19,426 MW served by wind generation. Most of wind and solar generation projects in ERCOT are interconnected at transmission level (i.e., 69 kV, 138 kV, or 345 kV). The best wind resource potential within the ERCOT region is in the Panhandle[1] and West Texas as well as along the Gulf Coast. These areas have an average wind generation capacity factor of about 43% (with consideration of curtailments due to transmission constraints). However, while the Gulf Coast area is generally more attractive due to higher energy prices as well as favorable wind generation pattern that's better correlated with ERCOT load peak, more wind generation has been built in the West Texas and

Texas Panhandle areas due to land availability and higher annual energy yield. Similar considerations apply to solar projects as well. It should be noted that the renewable resources in the ERCOT region are far from the load centers where most of conventional synchronous generation is also concentrated. This results in high power transfers over a long distance to deliver the renewable power to the load centers. In this environment, to ensure reliable operation of the ERCOT grid, adequate voltage support must be maintained, and associated voltage stability must be assessed. Wind and solar resources are connected to the grid through inverters. Inverter-based resources (IBRs) such as wind and solar, being low to zero marginal price resources, displace more expensive conventional synchronous generation in unit commitment and dispatch. As a consequence, the decline in system inertia and the corresponding increase rate of change of frequency (ROCOF) during events must also be assessed. Sufficient system reserves and flexibility are needed to manage the variability and uncertainty of renewable generation during real-time grid operations.

2.4 Capacity Adequacy and Wind Generation Resources

For the purpose of calculating the ERCOT generating reserve margin, ERCOT until recently was counting on 8.7% of installed wind capacity for all wind power plants regardless of their geographical location. This assumption was based on a 2007 study using loss of load probability methodology [10]. This capacity contribution number did not reflect subsequent increases in the number of wind generation projects and their geographic dispersion. Also, applying one capacity value for all wind power plants regardless of their location disregards the strong correlation between load patterns and wind generation in the coastal area.

In October 2014, a new methodology for calculating the capacity value of wind generation during peak load periods was approved. The new approach calculates average historical wind generation availability during the top 20 peak load hours, rather than the effective load carrying capability as determined by a loss of load probability studies. Wind generation availability is expressed as a percentage of installed wind capacity and evaluated over a multi-year period.[2] This methodology improves data and process transparency while also giving ERCOT the ability to update the wind capacity value on a more frequent basis [11].

This new approach also included determining the wind capacity values for all four seasons, as well as distinguishing between wind resources located in non-coastal and coastal wind regions, recognizing the differences in production patterns between these two areas.

---

[2]The final value is the average of the previous 10 eligible years of seasonal peak average values. Eligible years include 2009 through the most recent year for which data is available for the summer and winter peak load seasons. If the number of eligible years is less than 10, the average shall be based on the number of eligible years available.

By the end of 2015, capacity contribution from wind generation, based on 6 years of historical data, was 12% for non-coastal wind and 55% for coastal wind across the summer peak and 18% and 37%, respectively, across the winter peak.

3   ERCOT Energy and Ancillary Services Market

In 1999, Senate Bill 7 (SB7) restructured the Texas electricity market by unbundling the investor-owned utilities and creating retail customer choice in those areas and assigned ERCOT four primary responsibilities.

- System reliability—planning and operations
- Open access to transmission
- Retail switching process for customer choice
- Wholesale market settlement for electricity production

In 2001, ERCOT began its single control area operation and opened both its wholesale and retail electricity market to competition based on a zonal market structure. In the zonal market, the ERCOT region is divided into congestion management zones (CMZs) which are defined by the commercially significant constraints (CSCs).

In 2003, the Public Utility Commission of Texas (PUCT) ordered ERCOT to develop a nodal wholesale market design. The re-designed ERCOT grid consists of more than 4000 nodes, and it will replace the existing CMZs. The implementation of nodal market is expected to improve price signals, improve dispatch efficiencies, and direct the assignment of local congestions.

On December 1, 2010, ERCOT successfully launched the locational marginal pricing-based nodal market. The re-designed comprehensive nodal market includes congestion revenue right (CRR) auction market, a day-ahead market (DAM), reliability unit commitment (RUC), and real-time security-constrained economic dispatch (SCED).

A congestion revenue right (CRR) is a financial instrument that entitles the CRR owner to be charged or to receive compensation for congestion rents that arise in the day-ahead market (DAM) or in real time. Owning a CRR doesn't provide the CRR owner a right to receive or obligation to deliver the physical energy. CRRs are defined by a MW amount, settlement point of injection (source), and settlement point of withdrawal (sink).

There are two types of CRR ownership: point-to-point (PTP) obligations and point-to-point (PTP) options. The PTP obligation may result in either a payment or a charger for the CRR ownership, but the PTP options can only result in a payment for the CRR ownership. CRRs are auctioned by ERCOT monthly and annually, and the revenues collected from the auctions are returned to loads based on the load ratio share.

The day-ahead market (DAM) is a forward financial electricity market cleared in day-ahead. The DAM clearing process co-optimizes the energy offers and bids, ancillary services, and certain types of congestion revenue rights (CRRs) by maximizing system-wide economic benefits. The DAM clearing results include the unit commitments for resources with three-part supply offer, awards for energy offers and bids, awards for ancillary services, and awards for a certain type of CRRs. The DAM scheduling also complies with network security constraint in addition to the usual resource constraints.

The main purposes for the DAM are scheduling energy and ancillary services, providing price certainty and discovery for the next operating day. The reliability unit commitment (RUC) is a daily or hourly process conducted to ensure sufficient generation capacity is committed to reliably serve the forecasted ERCOT demand [6]. RUC is also used to monitor and ensure the transmission system security by performing the network security analysis (NSA). The DAM clearing is based on the voluntary energy offers and bids instead of the load forecast. The resources committed in the DAM may not be sufficient to meet the actual energy and ancillary service capacity requirements in real time. Hence, the RUC process is needed to procure enough resource capacity to meet load forecast in addition to ancillary service capacity requirement. The RUC process works like a bridge filling the capacity gap between the financial DAM and real time to ensure the reliable operation of the ERCOT market. There are three RUC processes used in the ERCOT nodal market:

- Day-ahead RUC (DRUC): DRUC runs once a day. It is used to determine if additional commitments are needed to be made for the next operating day.
- Hourly RUC (HRUC): HRUC process is executed every hour. It is used to fine-tuning the commitment decision made by DRUC based on the latest system condition.
- Weekly RUC (WRUC): WRUC process is an offline planning tool. Its study period is configurable and could be up to 1 week.

During real-time operations, the security-constrained economic dispatch (SCED) dispatches online generation resources based on their energy offer curves to match the total system demand provided by the EMS while observing resource ramping and transmission constraints. The SCED process produces the base point and locational marginal prices (LMPs) for each generating resource. ERCOT uses these base points to deploy various ancillary services such as regulation up, regulation down, responsive reserve, and non-spinning reserve services to control system frequency and solve potential reliability issues [10].

The ERCOT nodal market structure is illustrated in Fig. 1.11. The adjustment period is defined as the time between 1800 of day-ahead and the 60 minutes prior to the operating hour. The MIS denotes the market information system which is an electronic communication interface used by ERCOT to provide information to the public and market participants.

The two-settlement system [10] has been adopted for the ERCOT nodal market. The two-settlement system provides the ERCOT market participants with the option to participate in a forward market for energy. It consists of two markets, day-ahead

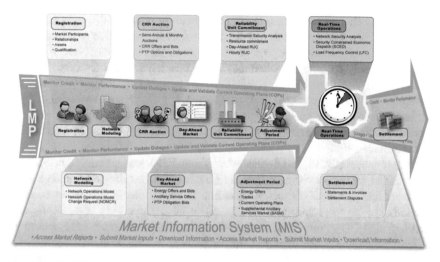

Fig. 1.11 ERCOT market

forward market and real-time balancing market, and it separates the settlements performed for each market.

The day-ahead market settlement is based on the scheduled hourly quantities and day-ahead hourly LMPs, and the real-time market settlement is based on the actual 15-minute quantity deviations from the day-ahead schedules priced at real-time LMPs.

3.1 Ancillary Services

Ancillary services in ERCOT currently include responsive reserve service, regulation service, non-spinning reserve service, black start, and emergency interruptible loads. The latter two services are used in emergency and system recovery conditions, while the former three services are used to balance net load[3] variability and support frequency after generation outages. These services are described in more details below.

Responsive reserve service (RRS) bundles two distinct functions within one service. This reserve is used for frequency containment, i.e., to arrest frequency decline after generator trip, and as a replacement reserve to restore the depleted responsive reserves and bring the frequency back to 60 Hz. Until recently (June 2015), ERCOT procured 2800 MW of RRS for every hour in a year. Of this amount, 50% could be provided by interruptible load resources (LRs) with automatic under-frequency relays. The relays are activated within 0.5 seconds, if system frequency

---

[3] Net load is defined as system load minus wind and solar power production.

drops to 59.7 Hz or lower. These interruptible loads providing RRS are usually large industrial loads. The remainder of RRS is provided by generation resources and is deployed autonomously through governor response (as containment reserve) and/or through security-constrained economic dispatch (as replacement reserve).

Recently, due to changing generation mix in ERCOT, the methodology for determining RRS has changed from procuring a constant amount of 2800 MW for all hours to determining necessary amounts of RRS dynamically based on expected system inertia conditions, and up to 60% of RRS could be provided by interruptible load resources.

A new subset of RRS, fast frequency response (FFR), was introduced in 2020. To be qualified for the provision of FFR, a resource should be able to be automatically deployed and provide its full response within 15 cycles after the frequency meets or drops below a preset threshold (59.85 Hz) or be deployed via a verbal dispatch instruction (VDI) within 10 minutes. FFR resources must sustain a full response for at least 15 minutes once deployed. In comparison to LRs, FFR will be deployed earlier and faster when a loss of generation event happens. Earlier response from FFR will aid in preserving LRs providing RRS for more severe events. With a trigger set at 59.85 Hz, FFR will deploy more frequently than LRs in the response to the under-frequency events. As FFR resources provide a frequency response, most frequency events will not trigger the LR deployment, thus preserving the frequency response capabilities from LRs to be available for the next severe event.

Regulation reserve service is a restoration service, used to restore frequency back to 60 Hz after a disturbance as well as to balance out intra 5-minute variability in net load. Resources providing regulation service respond to an automatic generation control (AGC) signal from ERCOT every 4 seconds. The regulation reserve require-ments are determined separately for regulation up and regulation down for each month, hours 1 through 24 (i.e., 24 values per month). The requirements are based on regulation service deployments for the same month in the previous 2 years and a certain percentile for the 5-minute net load variability for the same period. ERCOT also can increase regulation requirements based on the historic exhaustion of regulation reserves as well as adjustments to account for newly installed wind/ solar generation capacity that is not yet captured in the historic evaluation period [12].

Somewhat counterintuitively, regulation requirements in ERCOT are trending down with increasing installed wind capacity, as shown in Fig. 1.12. This can be explained by the increase of geographical dispersion of wind generation resources as well as the continuous fine-tuning of ancillary service methodology for determining the requirements.

Fast-responding regulation service (FRRS) was introduced in 2013 as a subset of regulation service to allow resources to provide fast frequency support to the system. Resources providing FRRS are required to respond to a separate AGC signal as well as to a local frequency trigger (currently set at 59.91 Hz). Each resource with FRRS obligation should provide its full response within 1 second.

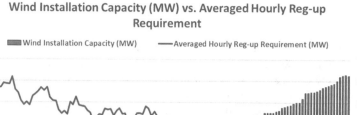

Fig. 1.12 Wind installed capacity vs. averaged hourly reg-up requirement

Non-spinning reserve service (NSRS) is a service to provide support within 30 minutes through online and/or offline resources. NSRS may be deployed to react to the loss of a generator, to compensate for net load forecast error, and to address the risk of large net load ramps or when low amount of generation capacity is available in SCED. Historically, the need for NSRS has occurred during hot or cold weather, during unexpected weather changes, or following large unit trips to replenish deployed reserves. The amount of NSRS is determined for every month in 4-hour blocks. The NSRS requirement is determined as a percentile of the 3-hour-ahead net load forecast errors plus a 75th percentile of the incremental intra-day outage in that block of hours from the same month of the previous 3 years. The percentile applied to net load forecast error is chosen based on the risk of high net load ramps in the period under evaluation. If the risk of the net load ramp is high, the NSRS requirement is based on the 95th percentile of the hourly net load forecast error; if the risk is low, the 85th percentile applies.

4   Reliability and Security of Grid Operations

As more renewable resources are interconnected to the ERCOT grid, the operators are faced with a set of challenges to maintain a satisfactory reliability and security for the grid operations, which are described in the following subsections.

4.1 Requirement for Primary Frequency Response

In ERCOT, all online generation resources, including renewable generation, are required to provide primary frequency response through governor or governor-like action to the changes in the system frequency outside of the narrow deadband of ±0.017 Hz [13]. Primary frequency response shall have a droop characteristic of a maximum of 5%. However, only resources that are awarded responsive reserve in the DAM are required to reserve a certain amount of capacity to meet their obligations. Other resources will provide primary frequency response only if they have available headroom.

4.2 ERCOT Frequency Performance

ERCOT's frequency performance is monitored by the North American Electric Reliability Corporation (NERC) through the Control Performance Standard (CPS1). CPS1 measures the quality of frequency control performance as follows [14]:

$$CPS1 = 100\% \left(2 - \frac{\overline{\Delta F_{1m}}^2}{\varepsilon^2} \right) \tag{1.1}$$

where ε (Hz) is determined by NERC and represents the historical performance of frequency control (for ERCOT, ε is 0.03 Hz) and $\overline{\Delta f_{1m}}$ is the 1-minute average of system frequency deviations sampled every 4 seconds. The 1-year average of CPS1 must be larger than 100% for system frequency to comply with the standard.

ERCOT's frequency performance is continuously improving over the past years (Fig. 1.13), despite a growing share of variable generation resources and reduction in regulation requirements (Fig. 1.12). This increase can be attributed to:

- Continuous fine-tuning of the procured AS amounts, based on historical data
- Primary frequency response requirements for all generators, including renewable generation
- Continuous performance monitoring by ERCOT after each significant frequency event
- Generators preparing for the implementation of BAL-TRE-001 standard that narrows governor deadband settings from 0.036 to 0.017 Hz

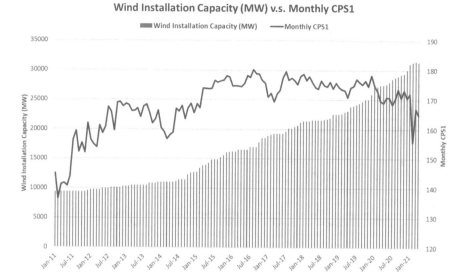

Fig. 1.13 Frequency performance in ERCOT from 2003 to 2015

4.3   System Inertia Trend

Rotating turbine generators and motors that are synchronously connected to the system store kinetic energy. In response to a sudden loss of generation, kinetic energy will automatically be extracted from the rotating synchronous machines causing the machines to slow down, and, as a consequence, system frequency is decaying. This is called inertial response. Inertial response provides an important contribution to reliability in the initial moments following a generation or load trip event determining the initial rate of change of frequency. However, non-synchronous resources like wind and solar generation do not contribute to the synchronous inertia.

The system inertia varies with the hours of day and seasons of year as the statuses of the synchronous generators change over the time. The system inertia is also correlated with the net load (which is the load minus the aggregated wind and solar generation). If there is an abundant wind or solar generation contemporaneous with low load conditions, wholesale energy market prices can be low or even negative. During these conditions, synchronous generators may be offline for economic reasons, which reduces system inertia. Table 1.1 shows the lowest inertia from 2013 to 2020. As the system inertia is declining, the grid could be susceptible to the reliability risk if there is lack of frequency responsive capacity at the time when the inertia is low.

Table 1.1 Lowest inertia in different years (GW·s)

Date and time	2013 3/10 3:00 a.m.	2014 3/30 3:00 a.m.	2015 11/25 2:00 a.m.	2016 4/10 2:00 a.m.	2017 10/27 4:00 a.m.	2018 11/03 3:30 a.m.	2019 3/27 1:00 a.m.	2020 05/01 2:00 a.m.
Min synch. inertia (GW·s)	132	135	152	143	130	128.8	134.5	131.1
System load at minimum synch. inertia (MW)	24,726	24,540	27,190	27,831	28,425	28,397	29,883	30,679
Non-synch. gen. In % of system load	31	34	42	47	54	53	50	57

4.4   Renewable Generation Forecasting

Renewable production potential (RPP) forecasting is essential to the integration of a large amount of renewable generation. ERCOT acquired its first wind forecast service in 2010 and added a second wind forecast service in 2017 to provide centralized renewable forecast for all renewable generation connected to ERCOT transmission grid. Both wind forecasting services utilize the telemetered site-specific meteorological information provided by renewable resources to generate wind forecast for each wind generation resource. Adding another wind forecasting service provided ERCOT's control room with an alternative wind forecast to enhance situational awareness and increased resiliency for the provision of this critical information. Along with the second provider, new capabilities of forecasting for extreme weather were also procured from both providers and made available to the control room. In addition to the mid-term wind generation forecast, which is delivered each hour for the next 168 hours, ERCOT also has an intra-hour wind forecast, which is delivered at a 5-minute resolution for the next 2 hours. The averaged performance, i.e., mean absolute percentage error (MAPE), of day-ahead and 1-hour-ahead mid-term wind forecasting in 2020 was 4.3% and 2.4%, respectively (Fig. 1.14).

In 2017, ERCOT started solar forecasting for solar generation connected to the transmission network with a size greater than 10 MW. The averaged performance of ERCOT's day-ahead and 1-hour-ahead mid-term solar forecasting in 2020 was 5.84% and 4.8%—see Fig. 1.15.

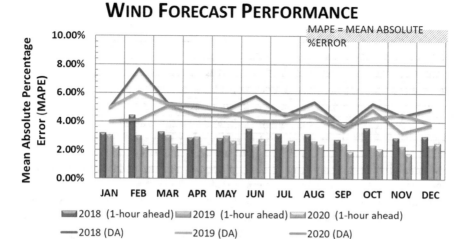

Fig. 1.14   Annual 1-hour-ahead wind forecast performance

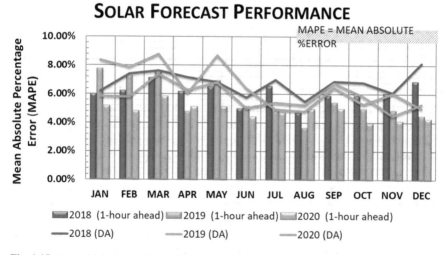

Fig. 1.15   Annual 1-hour-ahead solar forecast performance

4.5 Operations Analysis and Studies

To ensure reliable system operation in real time, the system operator periodically carries out studies and analysis starting from the day-ahead and going into real time and uses situational awareness tools to monitor system conditions.

4.5.1 Day-Ahead Analysis and Studies

Generic Transmission Limits (GTLs) for Day-Ahead Market—in addition to well-defined tools and process to manage the steady-state constraints, increasing stability-related constraints along with increasing IBRs connected to the electric grid also needs to be properly managed in operations. Generic transmission constraint (GTC) is a transmission constraint on one or more grouped transmission elements to manage stability-related constraints. A 2-day-ahead offline study is conducted to determine transfer limits based on the projected system condition on all existing GTCs. The identified GTC stability limits are used in the day-ahead market (DAM) and reliability unit commitment (RUC) processes and are updated in real time if necessary to maintain reliable transfers in real-time operation.

Next-Day Study (NDS)—this study typically is conducted at 2:30 p.m. that includes steady-state analysis, contingency analysis, and voltage stability analysis for the upcoming operating day to ensure that ERCOT has sufficient generation to reliably serve projected peak load while taking into account the DAM and DRUC results, applicable system outage information, and transmission line dynamic ratings if applicable. Mitigation plans may also be developed for any issues that may arise.

4.5.2 Real-Time Analysis and Studies

GAP Study—It is a supplement of the next-day study. The generation pattern for wind and solar is updated based on the latest current operating plans submitted by each resource. The generation input for GAP study is the result for the latest hourly reliability unit commitment run. This study, therefore, has a better outlook of the available generation compared to the NDS. GAP study uses the same process and NDS and includes steady-state, contingency, and voltage stability analysis.

Real-Time Voltage Security Assessment (VSA)—It is conducted every 10 minutes, using real-time system conditions, to determine the transfer limits of the concerned regions, based on steady-state voltage stability or system strength considerations (whichever is more binding). The determined transfer limits are implemented as GTLs to be used in real-time security-constrained economic dispatch.

Real-Time Transient Security Assessment (TSA) (Under Implementation)—Most dynamic stability-related GTCs are based on offline studies which generally include conservative assumptions to account for various potential real-time system conditions. ERCOT is planning to implement the transient security assessment in real-time operation in the near future to determine the limits based on the most recent real-time operations.

4.5.3 Situational Awareness Tools

Inertia Monitoring Tool—IBRs do not naturally contribute to system inertia. As a number of these resources continue to increase and displace synchronous generators in a power system's generation mix, the synchronous inertia will inevitably decline. In recent years, ERCOT carried out a number of dynamic studies, which are showing that the amount of responsive reserve service needed to arrest the frequency decline before triggering under-frequency load shedding (UFLS) after the largest generation loss (per NERC Standard BAL-003) depends on system inertia conditions. Additionally, ERCOT identified a critical inertia level to ensure sufficient system inertia can be maintained to minimize the risk of UFLS after generation loss. In 2013, ERCOT implemented the real-time inertia monitoring tool, which continuously calculates the current total inertia contribution of all online synchronous generators, based on the inertia parameters of individual units in the network model and the online status of the units in the energy management system. The tool also is calculating future inertia conditions for the next 168 hours on a rolling basis. This calculation is based on the unit commitment plans (for the next 168 hours) that every generator submits to ERCOT every hour. The tool then identifies any time periods where the expected system inertia is less than the critical level, in which case ERCOT system operators will follow procedures to deploy non-spinning reserve or start other synchronous generators in order to increase synchronous inertia online. Currently, ERCOT's all-time minimum inertia level is still about 30% higher than the determined critical inertia level.

Reserve Adequacy Tool—Another situational awareness tool developed at ERCOT is the reserve sufficiency tool. ERCOT publishes responsive reserve service amounts before the start of each year for every hour of the upcoming year. These requirements are based on historic inertia conditions from the past 2 years and vary by month and time of day. In real time, however, the system inertia may be different from what was expected based on historical data. The reserve sufficiency tool allows operators to verify whether available frequency responsive capacity, including procured reserves, is sufficient based on actual inertia and calculated future inertia conditions. The reserve sufficiency tool uses inertia information from the inertia monitoring tool and calculates the amount of reserve needed to protect against ERCOT's NERC BAL-003 defined generation loss event, using empirical equations obtained from the aforementioned series of dynamic studies. The tool then compares all available frequency responsive capacity with the amount of reserves needed. If available frequency responsive capacity, including reserves procured in the day-ahead market, is insufficient, ERCOT may open a supplemental ancillary services market and procure additional reserves as needed.

Physical Responsive Capability (PRC) Monitoring Tool—A monitoring tool used in operations as an indicator of overall system "health" in terms of available frequency responsive capability. PRC sums up all available frequency responsive capacity from online resources. If available PRC is below a certain threshold and cannot be restored to a higher level over a given time, ERCOT enters energy emergency alert, and the system operators will follow procedures and steps to recover the system back to normal conditions with sufficient PRC available.

Capacity Adequacy Tool—This tool is deployed at ERCOT to assess whether the generation capacity is sufficient to serve the forecasted load demand for the next 1 hour to 24 hours. It tracks and monitors the online generation capacity and reserves and offline generation capacity that can potentially be started within xxx hours. To account for the variability and uncertainty of intermittent resources, the tool also allows users to evaluate the impact of different levels of load and wind forecasting uncertainties over the system adequacy. Thus, operators have a better situation awareness for the future grid operation conditions and can effectively mitigate anticipated large net load ramp events.

4.5.4   Challenges for the Nearest Future

Increasing numbers of IBRs and declining online synchronous generation lead to lower system strength and higher voltage sensitivity. Voltage controls and coordination are increasingly difficult. Frequent adjustments of voltage profiles and reactive devices are expected due to the variation of renewable generation and load patterns. To address this challenge, ERCOT identified the need of reactive power control tool in operations to improve the control and coordination of various reactive devices, including generator voltage profile, switchable shunts, and FACTs devices for both real-time and day-ahead operations. Currently, there is no commercial tool available to meet this need, and ERCOT is in the process of developing and implementing such a tool in the future.

ERCOT has already implemented transient security assessment tool, TSAT, in operations in a study mode to assess the system dynamic stability on as-needed bases. However, due to the increasing dynamic stability challenges in the evolving ERCOT grid, ERCOT plans to implement real-time TSAT to assess real-time dynamic stability. Similar to the real-time VSAT, the real-time TSAT is expected to run automatically and with a certain frequency, e.g., every 10 minutes, to determine the dynamic stability limits on the defined transmission interfaces, and the identified limits will be managed through GTCs enforced in security-constrained economic dispatch.

As more wind and solar generation was built, ERCOT and its market participants had to introduce a number of changes to the market rules, ancillary services, and operation practices and develop new situational awareness tools and real-time studies to ensure continued system reliability. Based on ERCOT's experience, the following practices are recommended to ensure operational security with high IBR:

- Granular unit commitment and economic dispatch that takes into account the most up-to-date load and renewable forecast and generating unit status information.
- Some of the essential reliability requirements need to be implemented through interconnection requirements or grid codes [15].

- Ancillary services products need to be evolving with evolving generation mix, and faster more flexible service may be required as amounts of IBRs continue to increase.
- Granular telemetry from all generating resources is required to inform situational awareness tools and studies conducted in operations as well as ex-post analysis.
- Real-time situational awareness tools in the control room are essential for efficient and reliable operation.
- Conducting a number of studies and analysis in day ahead and real time allows maintaining reliable system operation, on the one hand, while simultaneously allowing more efficient operation under less conservative assumptions compared to pre-determined limits based on offline studies with worst-case scenario assumptions, on the other hand.

5 Conclusions

As new wind generation was first beginning to come online in ERCOT, its impact on the daily system operation was insignificant. However, as more wind generation was built, ERCOT and its market participants had to introduce a number of changes to the generation interconnection requirements, market rules and market design, ancillary services, and operation practices to ensure continued system reliability with increasing amounts of variable non-synchronous generation. ERCOT is continuously working on the further modification of ancillary services procurement methodology, tuning the AS amounts to satisfy ever-changing system needs. ERCOT is also researching possibilities to supplement diminishing inertial response from synchronous generation with the use of emerging technologies.

References

1. ERCOT, Capacity Demand Reserve Report, Available at http://www.ercot.com/content/wcm/lists/96607/CapacityDemandandReserveReport_May2016.xlsx
2. NERC, Interconnection Map, Available at http://www.nerc.com/AboutNERC/keyplayers/Pages/Regional-Entities.aspx
3. Bill Magness, "ERCOT in 2016: Transition and Continuity", presentation to Golf Coast Power Association, Houston, Nov 2015, Available at http://www.ercot.com/news/presentations/index.html
4. ERCOT News Release, "Energy use in ERCOT region grows 2.2 percent in 2015", January 15, 2016 http://www.ercot.com/news/press_releases/show/86617
5. U.S. Department of Energy, Texas Wind Resource Map and Potential Wind Capacity, http://apps2.eere.energy.gov/wind/windexchange/wind_resource_maps.asp?stateab=tx
6. Herman K. Trabish, U.S. Wind Industry Hits 70 GW Capacity Mark, Celebrates Tax Credit Extension, Utility Dive, December 22, 2015, http://www.utilitydive.com/news/us-wind-industry-hits-70-gw-capacity-mark-celebrates-tax-credit-extensio/411224/

7. ERCOT, Generation Interconnection Study Report, Available at http://www.ercot.com/gridinfo/resource
8. Global Wind Energy Council, Global Statistics in 2014, Available at http://www.gwec.net/global-figures/graphs/
9. Warren Lasher, The Competitive Renewable Energy Zones Process, presentation at the Quadrennial Energy Review Pubic Meeting on Local and Tribal Issues, August 11, 2014, Available at http://energy.gov/epsa/downloads/qer-public-meeting-santa-fe-nm-state-local-and-tribal-issues
10. Global energy Decisions, ERCOT Target Reserve Margin Analysis, January 2007, Available at http://www.ercot.com/content/meetings/gatf/keydocs/2007/20070112-GATF/ERCOT_Reserve_Margin_Analysis_Report.pdf
11. ERCOT, Nodal Protocol Revision Request 611: Modifications to CDR Wind Capacity Value,
12. ERCOT, Methodologies for Determining Ancillary Service Requirements, Available at http://www.ercot.com/mktrules/issues/NPRR611
13. NERC, BAL-001-TRE-1 – Primary Frequency Response in the ERCOT Region, Available at http://www.nerc.com/pa/Stand/Reliability%20Standards/BAL-001-TRE-1.pdf
14. NERC Resources Subcommittee, Balancing and frequency control, January 26, 2011, Available at http://www.nerc.com/docs/oc/rs/NERC
15. Hydro-Québec TransÉnergie, Transmission Provider Technical Requirements for the Connection of Power Plants to the Hydro-Québec Transmission System, February 2009, Available at http://www.hydroquebec.com/transenergie/fr/commerce/pdf/exigence_raccordement_fev_09_en.pdf

Chapter 2
Overview of Market Operation at ERCOT

1 Overview

Figure 2.1 depicts a summary of the components of the ERCOT wholesale market. This summary identifies elements of the market such as registration, network modeling, congestion revenue rights, day-ahead and real-time operations, settlements, and the market information system [1–4]. Though this summary shows the components of the ERCOT wholesale market as a sequential process, some of the components shown (such as registration) may occur at any time. Others (such as network modeling) are continuous functions. Registration is the process whereby data on market participants, and their assets, is received and maintained by ERCOT. All market participants are required to register and sign a standard form market participant agreement to participate in the ERCOT market.

Following registration, some specific market participants must demonstrate that they are capable of performing the functions for which they have registered as required by the protocols. This is demonstrated by completing qualification tests. These tests depend on the market participant's level of operational interaction with ERCOT.

Qualified scheduling entities (QSEs) submit bids and offers on behalf of resource entities (REs) or load serving entities (LSEs) such as retail electric providers (REPs). QSEs are required to successfully complete qualification tests. Qualification tests are intended to verify connectivity for communication with ERCOT as well as test the QSE's operational readiness. QSEs must prove their ability to submit information through the market manager user interface or through an automated programmatic interface (API). If the QSE represents resources, ERCOT must test their telemetry connectivity through the ERCOT wide-area network (WAN). Finally, if the QSE plans to provide ancillary services, they must qualify for each service they wish to provide. Resource entities must complete a resource asset registration form for each resource they own.

© The Author(s), under exclusive license to Springer Nature Switzerland AG 2023
P. Du, *Renewable Energy Integration for Bulk Power Systems*, Power Electronics
and Power Systems, https://doi.org/10.1007/978-3-031-28639-1_2

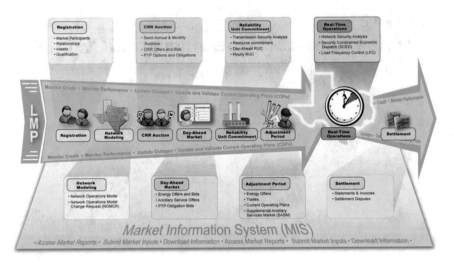

Fig. 2.1 Overview of market information system at ERCOT

In the nodal market, ERCOT will perform financial settlement with QSEs and congestion revenue rights (CRR)[1] account holders. Therefore, as part of their qualification process, new QSEs and CRR account holders are required to complete a credit application form and show that they satisfy ERCOT's creditworthiness requirements as defined in the ERCOT protocols. ERCOT will continually monitor the available credit limits of QSEs and CRR account holders and may ask for additional collateral based on their market activities.

1.1 Network Modeling

The network model management system (NMMS) is used to produce and maintain the network operations model (NOM), as shown in Fig. 2.2. All other network models, the planning and CRR models, are based on this model. This NMMS helps with the accuracy and consistency of all ERCOT network models. It provides a single repository for all network model information and a controlled process for tracking all network model topology changes submitted to ERCOT.

NOM is a detailed representation of ERCOT's transmission grid and all of its generation and load resources. The NOM is the foundation over which resource commitment and dispatch happens for real-time operations. The NOM will also be used to develop models for the purpose of system planning (i.e., the planning models). Based on the NOM, locational marginal prices used in day-ahead and

---

[1] A congestion revenue right (CRR) is a financial instrument that results in a charge or a payment to the owner, when the ERCOT transmission grid is congested in the day-ahead market (DAM).

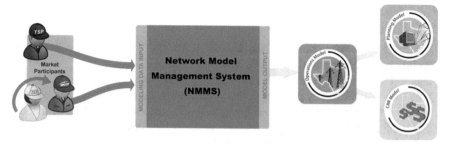

Fig. 2.2 Overview of network model management system (NMMS)

real-time operations will be calculated. It will also be used to develop models for conducting the congestion revenue rights market.

The NOM includes:

- Electrical busses—including configuration details.
- Transmission lines—including impedances and ratings. This may also involve dynamic ratings.
- Circuit breakers.
- Transformers.
- Disconnect switches.
- Reactive devices.
- Generation resources.
- Load resources.
- Protection schemes.
- Telemetry mapping details.

The modeling process depends on accurate inputs from TSPs and REs. TSPs provide changes to model data on transmission facilities, while REs must provide ERCOT with changes to generation or load resource information.

Planning models are used to help determine the needs and methods for handling system growth in the long term and identifying and studying new transmission or generation needs. ERCOT will develop models for annual planning purposes that contain, as much as practicable, information consistent with the NOM. Updates of the network planning model are coordinated with updates to the network operations model on the first of September each year and are released to market participants each year on October 15.

The CRR network model, based also on the NOM, is developed for use in CRR processes. The CRR model plays a key role in the running of the CRR auctions. ERCOT must update the CRR model and post it no later than 10 days before each monthly auction and no later than 20 days before each annual auction.

1.2   Day-Ahead Operations

Day-ahead operations consist of the day-ahead market (DAM) and the day-ahead reliability unit commitment (DRUC) processes. The DAM allows market partici- pants to bid to purchase or to offer to sell energy the day before an operating day. QSEs may also bid to buy point-to-point obligations in the DAM which are financial instruments that can be used to hedge against congestion in the real-time market. QSEs with resources may also offer to sell ancillary services in the DAM which are needed for the reliable operation of the grid.

The second part of day-ahead operations is the DRUC process through which ERCOT may bring resource capacity online needed for reliable operation of the system on the following day.

The DAM is a daily, co-optimized market for energy, ancillary services, and DAM point-to-point obligations. Participation in the DAM is voluntary for QSEs. QSEs do not have to represent generation or load to participate in the DAM. The DAM is financially binding, but not necessarily physically binding (except in the case of ancillary services). The process by which the DAM determines which bids and offers are awarded is said to be "co-optimized." "Co-optimized" means that the offers and bids for energy, point-to-point obligations, and ancillary services are simultaneously evaluated and cleared to ensure an optimal overall cost for all the products.

The reliability unit commitment or RUC process follows the completion of the DAM. ERCOT uses the RUC process to analyze the transmission system and determine if any additional resources need to be committed for the reliability of the following operating day. The day-ahead RUC process is the reliability portion of day-ahead operations. Its sole purpose is to ensure reliability for the ERCOT transmission grid—to see to it that there is enough resource capacity, in addition to ancillary services, committed in the right location for the next operating day.

Day-ahead reliability unit commitment, or DRUC, is also one part of the reliabil- ity unit commitment process. It is executed daily after the completion of the DAM. Day-ahead RUC may commit or de-commit resources for reliability purposes. The hourly RUC process is executed at the beginning of each operating period. In addition to assessing the need for additional resource capacity commitments needed for system reliability, the hourly RUC assesses whether or not resource de-commitment requests by QSEs may be granted.

1.3   Adjustment Period Operations

Upon completion of day-ahead operations, the adjustment period begins and con- tinues into the operating day. The duration of the adjustment period is defined for each operating hour and lasts from the end of day-ahead operations to the beginning of the operating period.

During the adjustment period, QSEs can submit, validate, and update offers, trades, and schedules. QSEs with resources may also request de-commitments of resources during this period for resources they had previously shown as committed by the QSEs. ERCOT, meanwhile, may take additional actions to assure that sufficient capacity and ancillary services will be available in real time. If ERCOT detects insufficiency of available ancillary service capacity, it may conduct a supplemental ancillary services market in the adjustment period.

1.4   Real-Time Operations

The purpose of real-time operations is to maintain the reliability of the ERCOT transmission grid to achieve this reliability goal; ERCOT must ensure that generation output and system demand are always matched and that the transmission system is operating within its established limits.

The process ERCOT uses to determine the optimal dispatch of resources in real time to meet system demand is referred to as the security-constrained economic dispatch (SCED) process. This process calculates resource-specific base points, locational marginal pricing (LMP), and shadow prices every 5 minutes.

Between SCED runs, there is another process called load frequency control (LFC). While SCED dispatches resources based on economics, LFC is charged with the ultimate goal of maintaining reliability of the power system by matching system generation and system demand. When system generation and demand are in perfect balance, the system frequency is 60 cycles per second or 60 Hz. As ERCOT dispatches generation to follow the system demand, the frequency of the system will deviate slightly from 60 Hz. The frequency is higher when there is more generation than demand and lower when there is more demand than generation.

LFC deploys resources providing regulation reserve service up or down every few seconds in response to frequency deviations from 60 Hz by the amount needed to correct the system frequency to 60 Hz. In doing so, LFC does not consider costs. It simply allocates the amount of regulation needed to correct system frequency proportionally across all QSEs providing regulation reserve.

2   Day-Ahead Market (DAM)

2.1   DAM Overview

To buy or sell products in the nodal market, an entity need to submit the "bids" or "offers." A bid is a proposal to buy a product. When a QSE submits an energy bid into the DAM, it is expressing its willingness to buy energy at a specific location or settlement point and at or below a specified price. On the other hand, an offer is a proposal to sell. An energy offer in the DAM by a QSE would state the QSE's

Fig. 2.3 Products bought and sold in the DAM

Fig. 2.4 Co-optimized
clearing process

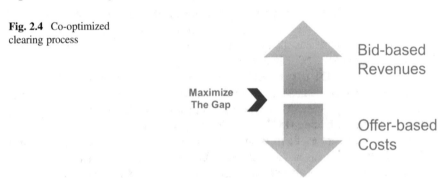

willingness to sell energy at a specific location or settlement point at or above a specified price.

As indicated in Fig. 2.3, QSEs have three options to submit bids or offers.

- First, QSEs may submit bids to buy or offers to sell energy into the DAM.
- Second, QSEs may bid to buy DAM point-to-point obligations. These are financial instruments that can be used to hedge against differences in prices between the DAM and the real-time market.
- Third, QSEs may also offer to sell ancillary services to ERCOT.

The DAM uses a co-optimized market clearing engine to award energy bids and offers, point-to-point obligation bids, and ancillary service offers submitted by QSEs. As the DAM runs, it will clear offers, which typically incur costs for the system. At the same time, it will clear bids which typically bring revenue into the system.

The objective of the DAM is to maximize bid-based revenues minus offer-based costs. Another way to look at this process is to maximize the gap between the total bid-based revenues and the total offer-based costs, as shown in Fig. 2.4. The costs include energy and ancillary services. Maximizing the gap between total revenue and total costs allows us to choose the most cost-effective way of using resources for energy and ancillary services.

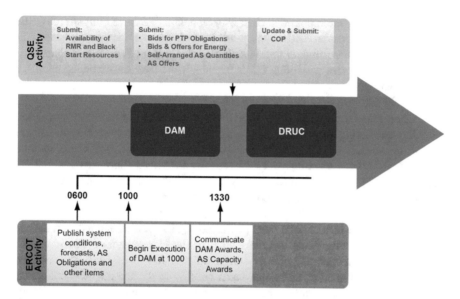

Fig. 2.5 Timeline of DAM activities

Participation in the DAM is voluntary, with some exceptions, but the outcome of the DAM is financially binding on those who participate. This means that QSEs are not obligated to submit offers and bids in the DAM. It also means that if a QSE makes offers and bids into the DAM, and these transactions are awarded, the submitting QSE is financially bound to any obligations resulting from the clearing of the market. Therefore, if a QSE's offer to sell energy in the DAM is awarded, the award becomes a financial obligation which may or may not be backed up with physical generation in real time.

On the other hand, awards for ancillary services are physically binding in that they result in an obligation for the QSE to make a specific resource capacity available for the deployment of ancillary services by ERCOT in real time.

DAM activities must occur during a specified timeline. Figure 2.5 is a representation of the DAM timeline that shows those times.

As indicated, the DAM timeline starts each day at 0600 (6:00 a.m.) with the publication of market information by ERCOT and ends with the publication of market results no later than 1330 (1:30 p.m.)

By 6:00 a.m., day ahead, ERCOT must publish information to market participants for the operating day. The information to be posted includes ancillary service obligations and system conditions such as an updated network model, forecasts, load profiles, and loss factors. Prior to this, QSEs representing reliability must-run units and black start resources must let ERCOT know of their availability status.

Prior to 10:00 a.m., QSEs may submit any energy bids, energy offers, point-to-point obligation bids, and ancillary service offers to be evaluated in the DAM. Although not used in the DAM, QSEs may also submit trades, self-schedules, and DC tie schedules.

Using the information posted at 6:00 a.m. on ancillary service obligations, QSEs representing load must let ERCOT know what part of their obligations they intend to self-arrange by either providing the ancillary services from their own resources or through bilateral trades with other QSEs by submitting their self-arranged ancillary service quantities.

At 10:00 a.m., ERCOT executes the DAM.

No later than 01:30 p.m., ERCOT will notify QSEs of the results of the DAM including notification of awarded energy offers and bids, ancillary service offers, and point-to-point obligation bids. Other results of the DAM posted by ERCOT include DAM LMPs, market clearing prices for capacity for the ancillary services market, aggregated quantities and offer curves for ancillary services, and aggregated amounts of energy bought and sold at each settlement point for each hour of the operating day.

Once the DAM results have been published, QSEs with resources must update their current operating plans to reflect ancillary service awards. QSEs may also update their current operating plan or COP to reflect DAM energy awards. Since the DAM is a financial market, an awarded energy offer results in a financial obligation and does not necessarily have to be backed up by a physical resource. If an energy offer that has been awarded is not backed up by generation in the real-time market, the QSE will be charged for the amount of energy awarded at real-time prices.

QSEs may also continue to submit trades and schedules through the adjustment period.

2.2 ERCOT and QSE Activities in DAM

In this section, we will describe the activities to be performed by ERCOT and QSEs at various times during the DAM timeline.

2.2.1 Ancillary Service Plan Posted in DAM

As discussed, ERCOT must publish system conditions, load forecasts, load profiles, ancillary service obligations, and loss factors.

To begin, we'll first review the ancillary service obligations in more detail, specifically looking at the relationship between the ancillary service plan that ERCOT creates and the resulting obligation of a QSE.

The ancillary service plan identifies the types and amount of system-wide ancillary services for each hour of the operating day that ERCOT determines necessary for reliable operations of the grid and to meet NERC standards for reliability. The MW of each ancillary service required may vary from hour to hour depending on ERCOT system conditions.

Using a methodology approved by the ERCOT board at least on an annual basis, ERCOT determines the minimum quantity of each ancillary service required for reliable operations. On a daily basis, ERCOT develops the ancillary service plan which is a plan that identifies the types and amount of ancillary service necessary for each hour of the operating day. ERCOT publishes the ancillary service plan no later than 6:00 a.m. of the day ahead.

After ERCOT determines the ancillary service plan, it then assigns ancillary service obligations for each ancillary service type for each hour of the operating day to each load serving entity (LSE) based on its load ratio share. ERCOT then aggregates each LSE's assigned obligations by QSE resulting in ancillary service obligations by QSE, by hour, and by ancillary service type. This information is posted by 6:00 a.m. of the operating day so that each QSE representing load can view its obligation.

For each hour of the operating day, the values of load ratio share used in ancillary service obligation assignments for this posting are based on each LSE's load ratio share for the same hour and day of the week for the most recent day for which initial settlement statements are available. Based on initial settlements for the real-time market being published 10 days after the operating day, the most recent statement available for the same day of the week as the operating day is from 14 days prior to the operating day. Although ancillary service obligations for a given operating day are assigned to QSEs based on load data from a prior operating day, these obligations are adjusted based on the most current real-time settlement and resettlement data for the operating day.

Once the QSE is aware of its ancillary service obligations, it has several options. The QSE may either choose to self-arrange the resources that can provide ancillary services or have ERCOT purchase ancillary services in the DAM on its behalf. The QSE may use either option to meet all or part of its ancillary service obligation. If the QSE chooses to self-arrange all or part of its obligation, it may do so by one of two ways. If the QSE also represents resources that can provide ancillary services, it may choose to self-provide the ancillary services. It may also choose to arrange to purchase ancillary services from other QSEs in the form of a trade and report the trade to ERCOT.

Once ERCOT knows the quantity of ancillary services being self-arranged by all QSEs with obligations, it is able to calculate the amount of each ancillary service that needs to be procured in the DAM. ERCOT procures the amount of ancillary services that are not self-arranged in the DAM using offers for ancillary services submitted by QSEs representing resources. Once the DAM is completed and these QSEs are notified of their awarded ancillary service offers, these QSEs will be able to determine their ancillary service supply responsibility which is the net amount of ancillary service capacity that the QSE is responsible to deliver by hour and service type, from resources represented by the QSE.

Once a QSE's representing resources providing ancillary services knows its ancillary service supply responsibility, it would then proceed to disaggregate the QSE-level responsibility into ancillary service quantities in MW to be provided by each resource it represents. The MW of each ancillary service that each resource is obligated to provide is referred to as the ancillary service resource responsibility.

Once QSEs determine the ancillary service resource responsibility of each of their resources, they must reflect these amounts in the current operating plans of each resource submitted to ERCOT to notify ERCOT of the resources that will be used to supply the QSE's ancillary service supply responsibility.

This information containing the hours for which the resource will provide ancillary services, type of ancillary service, and MW amount of capacity must be submitted to ERCOT no later than 1430 or 2:30 p.m. in the day ahead or the execution of the day-ahead reliability unit commitment process.

2.2.2 QSE Activities

Figure 2.6 depicts a more detailed breakdown of QSE activities in the DAM. Notice that each group of activities is tied to specific times. We will explore each activity in detail.

2.2.2.1 Report Availability of Must-Run and Black Start Resources

As previously mentioned, prior to 6:00 a.m., QSEs representing reliability must-run or black start resources must submit information to ERCOT indicating availability of these resources for the operating day and any other information that ERCOT may need to evaluate use of these units as per the applicable agreements with ERCOT.

2.2.2.2 Submit Bids for PTP Obligations

Any bids or offers that the QSEs want to have considered in the DAM must be submitted prior to 10:00 a.m. This is the market's opportunity to participate in the

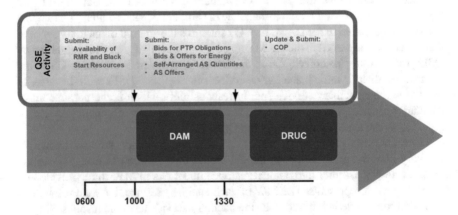

Fig. 2.6 QSE activities

Fig. 2.7 Schematic for PTP
obligation

forward energy market as well as to sell ancillary services. QSEs may also bid to buy PTP obligations in the DAM. PTP obligations are financial instruments that may be used as hedges against congestion costs in the real-time market. They are defined between a source settlement point and a sink settlement point. Although DAM PTP obligations are instruments for hedging against congestion costs, their use and financial settlement in the ERCOT market are very different from CRRs.

CRRs in the ERCOT market are settled using DAM prices. Therefore, they may be used as hedges against congestion costs in the DAM. CRRs do not offer any protection from the difference in congestion costs between the DAM and real-time operations. DAM PTP obligations may be used to hedge the additional risk of differences in congestion costs between day-ahead and real-time operations.

The purchase price of a DAM PTP obligation is the difference in the DAM settlement point prices between the sources and sink settlement points. Ownership of these PTP obligations entitles the owner to collect the congestion rent collected when there is congestion on the ERCOT grid in real time. Therefore, if a QSE's PTP obligation bid is awarded in the DAM, it pays for the awarded bid based on the difference in the DAM settlement point prices between the source and sink. The QSE will then be settled in real time based on the difference in the real-time market settlement point prices between the source and sink.

To formulate a DAM point-to-point obligation bid, the QSE must specify the two settlement points that define the obligation they wish to buy—see Fig. 2.7. They also must stipulate how many MWs they want and the sink-minus-source price they are willing to pay. A QSE may buy these financial instruments to hedge their real-time congestion cost exposure. However, a QSE does not need physical load or generation in order to buy these instruments. They may simply buy point-to-point obligations speculatively as an investment.

2.2.2.3 Submit Three-Part Supply Offers

Bids and offers for energy must be submitted by 10 a.m. as well. Energy offers submitted by QSEs may take two forms: three-part supply offers and day-ahead energy-only offers. A three-part supply offer is an offer to start and sell energy from a generation resource represented by the QSE. A three-part supply offer is composed of:

- Offer to sell energy from a generation resource
- Must specify:

 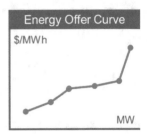

Fig. 2.8 Illustration of three-part offer

- Resource name
- First and last hour
- Expiration date and time

As the name implies, there are three parts to the offer, a startup offer, a minimum-energy offer, and an energy offer curve, as shown in Fig. 2.8. It is important to note that the QSE must specify the individual resource they are offering, the operating day for which the resource is offered, and the expiration date and time of the offer.

The startup offer is the total cost to startup in dollars per start. It represents all costs incurred by a generation resource in starting up and reaching the low sustained limit. The startup offer is actually three separate offers: hot start, intermediate start, and cold start. The minimum-energy offer is the cost per megawatt per hour to operate at the low sustained limit. The startup offer and minimum-energy offer can only be submitted as part of the three-part supply offer.

The protocols define maximum values or caps for the startup and minimum-energy offers submitted by QSEs. Each resource's startup offer and minimum-energy offer must be equal to or less than 200% of the resource category generic startup cost or minimum-energy cost for that type of resource. Generic caps on startup offers and minimum-energy offer based on the technology type of resources are defined in the protocols.

When the QSE submits a three-part supply offer, they must provide the percentages of fuel index price and fuel oil price for generation at LSL. ERCOT uses these percentages to determine the resource category generic minimum-energy cost.

QSEs may submit verifiable cost data to ERCOT that will enable ERCOT to determine verifiable startup and minimum-energy costs for each resource. Once verifiable startup and minimum-energy costs are approved by ERCOT, the QSE may submit offers up to 200% of the verifiable costs.

The energy offer curve is a proposal to sell energy at a settlement point. It must be submitted as a series of monotonically increasing prices with increasing MW levels of dispatch representing the price at which the QSE is willing to be dispatched at various MW operating levels. It is resource-specific and can be submitted independent of the startup offer and minimum-energy offer.

An energy offer curve must include the following: a monotonically increasing offer curve. The actual energy offer curve is submitted using price and quantity pairs,

Table 2.1 Consideration of three-part supply offer curve in different processes

	Three-Part Supply Offer		
	Startup Offer	Minimum Energy Offer	Energy Offer Curve
Day-Ahead Market	✓	✓	✓
RUC-Commitment	✓	✓	
Real-Time Dispatch			✓

limited to a maximum of ten. The minimum quantity that may be offered is 1 megawatt, although the incremental steps may be in tenths of megawatts. Prices must be between negative \$250/MWh and the system-wide offer cap of \$9000/MWh.

When energy and AS offers are made from the same resource, the linked offers must be designated as being inclusive or exclusive of each other. Inclusive offers allow the day-ahead market to clear both energy and ancillary services from the same resource. Exclusive offers allow the day-ahead market to clear either energy or ancillary services, but not both.

ERCOT also needs to know the percentage of fuel mix used by each resource. As part of the energy offer curve, QSE must submit the percentage fuel mix for generation above the resource's LSL. These percentages are used to determine what percentage the fuel index price or fuel oil price ERCOT will apply in calculating mitigated offer caps and mitigated offer floors in real-time operations.

Table 2.1 shows how a three-part supply offer is considered in the DAM, in the reliability unit commitment process, and in real-time operations. For example, when a QSE submits a three-part supply offer for a resource, that QSE is offering energy in the DAM. The day-ahead process will evaluate all three parts of the offer in its co-optimization engine. The offer also represents a willingness on the part of the resource to be committed by the reliability unit commitment or RUC process. The RUC process will evaluate each resource's startup costs and minimum-energy offer in determining the cost of commitment for each resource.

When a QSE submits an energy offer curve for a resource, without submitting a startup offer and a minimum-energy offer, it is offering energy in the DAM at a certain price as well as expressing the prices at which it is willing to be dispatched in real time.

2.2.2.4 Submit Energy Bids and Energy-Only Offers

There are certain types of energy bids and offers that are *only* applicable to the DAM. These are called DAM energy bids and DAM energy-only offers.

Day-ahead energy-only offers are not associated with physical resources. They represent a willingness to sell energy at specific settlement point at or above a specified price. Because energy-only offers are not resource-specific, they may be submitted at any settlement point including load zones and hubs and are often referred to as virtual offers. If awarded in the DAM, the QSE ends up with a financial obligation for the awarded quantity of energy at the real-time settlement point price.

If a QSE's offer to sell energy in the DAM is awarded, the QSE is paid for its awarded energy-only offers at DAM prices for the settlement point prices at which the offer was made. This DAM sale results in a financial obligation for the same amount of energy at the same settlement points settled using real-time settlement point prices. This means that the QSE essentially buys back the energy at the real-time settlement point price.

Energy bids submitted by QSEs represent willingness to buy energy in the DAM at or below a specified price at specific settlement points. The DAM allows QSEs to participate in this centralized forward energy market, facilitated by ERCOT, in which they will be able to declare the maximum MW amount they are willing to buy and the maximum price they are willing to pay. This is the price certainty that forward markets will offer QSEs.

The DAM allows QSEs with load obligations and the LSEs they represent to purchase energy needed to serve their load at a price they are willing to pay. The DAM is the final opportunity for these entities to purchase energy before real time. In conjunction with bilateral trades that may be arranged many weeks or years earlier, these forward purchases allow those with exposure to real-time prices to minimize their exposure. The amount of load for which QSEs with load will be charged for at real-time prices is the difference between the QSEs real-time load obligation and the amount covered by bilateral trades and any DAM purchases.

Energy bids can also be "virtual bids," meaning that QSEs do not have to have physical load to submit energy bids in the DAM. If a QSE's bid to buy energy in the DAM is awarded, the QSE is charged for its awarded energy bid at DAM prices for the settlement point prices at which the bid was made. This results in a financial credit for the same amount of energy at the same settlement points at real-time settlement point prices. This means that the QSE essentially sells back the energy or that the QSE will be paid for the energy at the real-time settlement point price.

The submittal criteria are similar for both energy-only offers and energy bids.

A QSE must indicate its name—as the selling QSE for the offer and as the buying QSE for the bid—and identify the settlement point of the offer or bid. The QSE must specify the first and last hour of the operating day for which the offer or bid is being submitted and the expiration time and date of the offer or bid. The offer or bid must also be identified as a fixed quantity block, variable quantity block, or a curve. Just as in the energy offer curves discussed earlier, energy-only offers and energy bids may be submitted as curves using price quantity pairs. Another way QSEs may specify prices at which they are willing to buy or sell is through block bids or offers where the QSE states the price it is willing to buy or sell energy for multiple hour blocks. Block offers may be submitted as either a fixed quantity block offer or a variable quantity block offer.

2.2.2.5   Submit Self-Arranged Ancillary Service

QSEs may either self-arrange the resources that will provide ancillary services on their behalf or allow ERCOT to procure the ancillary services needed to meet their obligations in the DAM.

A QSE can choose to self-arrange all or part of its ancillary service obligation by self-providing the capacity from resources it represents or by arranging for the delivery of the ancillary services through ancillary service trades with other QSEs. Regardless of how QSEs choose to self-arrange their ancillary services, ERCOT only needs to know each QSE's total self-arranged quantities so it can determine how much of the system-wide ancillary service requirements remain to be procured in the DAM.

As previously mentioned, the quantity of megawatts that a QSE chooses to self-arrange must be submitted to ERCOT by 10:00 a.m., prior to the execution of the DAM. This enables ERCOT to determine how much remaining ancillary service capacity needs to be obtained through the DAM.

Once the results of the DAM are posted and prior to 1430 or 02:30 p.m., each QSE must notify ERCOT about how it intends to supply its self-arranged ancillary services.

For the self-arranged amount that the QSE will supply using its own resources, this notification consists of the QSE updating its current operating plan.

2.2.2.6   Submit Ancillary Service Offers

ERCOT will procure ancillary services using the available ancillary service offers submitted by QSEs before the DAM is executed. An ancillary service offer is an offer to supply one or more ancillary service capacities in the DAM or a supplemental ancillary services market—see Table 2.2.

Ancillary service offers, like day-ahead energy offer curves and bids, also need to indicate whether the offer is a fixed quantity block or a variable quantity block. For ancillary service offers, a load resource can submit both a fixed quantity and a variable quantity block. A generation resource can only submit a variable quantity block.

Fixed quantity offers for ancillary services can only be submitted by load resources. A fixed quantity offer is made up of a single price and a single quantity for each hour of the offer. Just as in multi-hour block offers for energy discussed earlier, multi-hour block offers for ancillary services must be awarded for the entire MW quantity and for all hours of the offer or none at all.

Table 2.2 Ancillary service
offers

Type of AS	Offer
Reg Up	Price / MW
Reg Down	Price / MW
Responsive Reserve	Price / MW
Non-Spin	Price / MW

Variable quantity block offers for ancillary services can be submitted by generation and load resources. A variable quantity block is specified with a single price and single "up to" quantity (in MW) contingent on the purchase of all hours offered in that block.

All ancillary service offers must be submitted with prices that are less than the system-wide offer cap. The minimum amount per resource for each ancillary service product that may be offered is 0.1 MW. A resource may offer more than one ancillary service.

2.2.2.7 Update and Submit Current Operating Plan

Once the DAM has been executed and its results have been posted by ERCOT, QSEs with resources must update their current operating plans to reflect any awarded ancillary service or energy offers. These updates must occur prior to 1430 or 2:30 p. m. which is when ERCOT initiates the day-ahead reliability unit commitment process.

Recall that the DAM is a financial market and that energy awards resulting from it are only financially binding. By updating its current operating plan to meet the day-ahead energy awards, the QSE is making the financial commitments into a physical commitment to provide energy in real time.

As the name implies, the current operating plan or COP is a plan for each resource submitted by the QSE. The current operating plan indicates a resource's anticipated operating conditions and provides information on the status, limits, and plans of the QSE to provide energy and/or ancillary services from each of the resources it represents.

In day-ahead operations, the current operating plan provides information into the DAM and the day-ahead reliability unit commitment process. The COP is also used as an input into the hourly reliability unit commitment processes executed on an hourly basis at the end of the adjustment period for each operating hour.

During real-time operations, ERCOT no longer uses the plans submitted by QSEs in their current operating plans. Instead, ERCOT switches to using real-time telemetry for all information regarding the operating conditions of resources.

A QSE's current operating plan includes operating conditions for each of its resources for each hour in the next seven operating days. To ensure that the information included in the plan is as reliable as possible, it can be modified until the end of the adjustment period.

Each QSE that represents a resource shall update its COP reflecting changes in the availability of any resource as soon as reasonably practicable, but in no event later than 60 minutes after the event that caused the change.

2.3 DAM Engine

The purpose of the day-ahead market is to economically and simultaneously clear offers and bids submitted by QSEs for the various products such as energy, ancillary services, and DAM point-to-point obligations. The DAM engine simultaneously considers all the bids and offers and determines a solution that maximizes bid-based revenues minus offer-based costs, subject to generation and transmission constraints and the procurement of ancillary services. The results of the DAM are awarded offers and bids for energy, awarded bids for DAM point-to-point obligations, and awarded ancillary service offers.

In this case, bid-based revenues refer to revenues from day-ahead market energy bids and point-to-point obligation bids. The offer-based costs refer to costs from awarded three-part supply offers, energy offer curves, day-ahead market energy-only offers, CRR offers, and ancillary service offers.

The DAM process must honor security constraints such as transmission constraints as reflected in the updated day-ahead network model and resource constraints for resources associated with three-part supply offers submitted by QSEs. Other constraints that must be considered include linked offers for energy and ancillary services as well as block offers and bids.

Remember that the DAM is not working under the constraint of serving forecasted load of the ERCOT system. The DAM is voluntary, and all loads are not required to participate.

In executing the DAM to economically and simultaneously clear offers and bids, ERCOT must not only use an accurate network model, but it must also model the offers and bids received from QSEs at settlement points throughout the system. For example, ERCOT must determine the appropriate load distributions to allocate energy offers and energy bids submitted at load zones. Similarly, DAM point-to-point obligation bids with source and sink at load zones must also be modeled as energy injections and withdrawals across electrical buses in the designated load zones.

Load distribution factors map out the location or distribution of load across the electrical buses in the network model. The distribution of load across a network

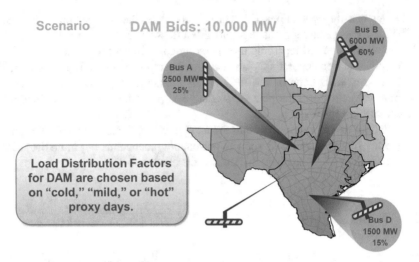

Fig. 2.9 Illustration of load distribution factors

model will determine the power flow in the modeled ERCOT system. The default load distribution factors used in clearing the DAM for a given hour come from the state estimator hourly distribution for the same day of the week 7 days prior. ERCOT, in its sole discretion, may change this proxy day from which the load distribution is selected to another proxy day reasonably reflecting the anticipated distribution in the operating day for reasons such as anticipated weather events or holidays.

Let's consider a scenario where the total number of energy bids at a load zone equals 10,000 MW, as shown in Fig. 2.9. In this case, the load zone has four electrical buses, and their hourly load distribution factors for the same day of the week prior from the state estimator are 25%, 60%, 0%, and 15%, respectively. The 10,000 MW of energy bids will be distributed to each electrical bus based on the load distribution factor for each bus. In addition, once the DAM is executed and day-ahead LMPs are determined at each electrical bus, the same load distribution factors are used to calculate the load zone settlement point price. The settlement point price for a load zone is the load-weighted average of the LMPs at each electrical bus in that load zone.

HUBs also have distribution factors, but they are provided in the ERCOT protocols. The HUB distribution factors are specified in the definitions of each HUB. The DAM process then uses this information to allocate bids and offers at a HUB to the appropriate buses in that HUB. This allocation will also influence the day-ahead LMPs and congestion costs.

We will use a very simplistic scenario shown in Fig. 2.10 to look at the details of how the co-optimization of DAM works.

In this scenario, ERCOT must procure 1 MW of regulation up and 1 MW of responsive reserve. Each of QSEs A and B has provided offers for 2 MW, which can

Scenario   For HE 1300 ERCOT needs to procure:

- 1 MW of Regulation Up (RegUp) and;
- 1 MW of Responsive Reserves (RRS)

Three QSEs submitted Offers and Bids as follows:

	Bids		Offers			
	MW	Energy	MW	Energy	RegUp	RRS
QSE A			2	$25.00	$10.00	$5.00
QSE B			2	$30.00	$11.00	$9.00
QSE C	1	$50.00				

Fig. 2.10 An example of DAM co-optimization

be cleared for either ancillary services or energy. QSE *C* has submitted a bid for 1 MW of energy.

DAM objective is to maximize bid-based revenue minus offer-based cost, while the bid-based revenue is the sum of bid price multiplied by cleared bid quantity, and the offer-based cost is the sum of offer price multiplied by cleared offer quantity.

We can determine the most desirable outcome by looking at the different combinations of cleared bids and offers. In our scenario, the bid side is easy since we have only the energy bid from QSE *C*. Thus, in all cases, our bid-based revenue is $50, since that is the bid price submitted by QSE *C*. What we will do, then, is look at the various combinations of cleared offers and calculate bid-based revenues minus offer-based costs for each. The combination that gives us the highest result is the winner.

As we can see from Fig. 2.11, Case 2 gives us the highest result. In other words, Case 2 maximizes bid-based revenues minus offer-based costs. Consequently, the DAM will procure 1 MW of energy and 1 MW of responsive reserve from QSE *A* and 1 MW of regulation up from QSE *B*. However, one question still remains: what will the prices be?

As we determine the energy price, the market is based on locational marginal pricing. As such, the energy price will be the cost to serve an additional increment of demand. Therefore, to determine the energy price for the above scenario, we must look at how the DAM would clear an additional MW of energy. However, another MW of energy cannot be taken from QSE *A*. They only offered 2 MW, and we have already used 1 MW for energy and 1 MW for responsive reserve service.

Alternatively, the additional energy can be provided from QSE *B* at $30. That would increase the offer-based cost by $30. What would happen if we moved the responsive reserve award to QSE *B* at $9? The offer-based cost of responsive reserve would increase by $4, but it would allow us to get the additional MW of energy from QSE *A* at $25. The net result is that the offer-based cost would increase by $25 + $4 or $29, less than the cost of getting the additional energy from QSE *B*. This combination gives us the most desirable outcome. Since our offer-based cost of an

Scenario

Case 1

	Energy	RegUp	RRS
QSE A	$25.00	$10.00	
QSE B			$9.00

Bid-based Revenues	—	Offer-based Costs
$50 − $25 − $10 − $9 = $6		

Case 2

	Energy	RegUp	RRS
QSE A	$25.00		$5.00
QSE B		$11.00	

Bid-based Revenues	—	Offer-based Costs
$50 − $25 − $11 − $5 = $9		

Case 3

	Energy	RegUp	RRS
QSE A		$10.00	$5.00
QSE B	$30.00		

Bid-based Revenues	—	Offer-based Costs
$50 − $30 − $10 − $5 = $5		

Fig. 2.11 Scenarios for DAM

Offers

	MW	Energy	RegUp	RRS
QSE A	2	$25.00	$10.00	$5.00
QSE B	2	$30.00	$11.00	$9.00

Clearing Additional MW of Energy

	Energy	RegUp	RRS
QSE A	1MW @$25.00		$5.00
QSE B	1MW @$30.00	$11.00	

→ Increases cost by $30

	Energy	RegUp	RRS
QSE A	2MW @$25.00		
QSE B		$11.00	$9.00

→ Increases cost by $29

Fig. 2.12 Energy prices for DAM

additional MW of energy is $29, our energy price (locational marginal price) is $29/MWh as shown in Fig. 2.12.

In a similar fashion, the price for each ancillary service will be the cost to clear an additional increment of the particular service. To clear an additional MW of responsive reserve, we would procure it from QSE *B*, so the responsive reserve price is $9. An additional MW of regulation up would have to come from QSE *B* as well, so the regulation up price is $11/MW—see Fig. 2.13.

Scenario

Offers	MW	Energy	RegUp	RRS
QSE A	2	$25.00	$10.00	$5.00
QSE B	2	$30.00	$11.00	$9.00

Clearing Additional Ancillary Services

	Energy	RegUp	RRS
QSE A	$25.00		1 MW @ $5.00
QSE B		$11.00	1 MW @ $9.00

	Energy	RegUp	RRS
QSE A	$25.00		$5.00
QSE B		2 MW @ $11.00	

Fig. 2.13 Ancillary service prices for DAM

In summary, QSE *A* sells 1 MW of energy with LMP of $29/MWh and 1 MW of RRS with the market clearing prices for capacity (MCPC) of $9/MW. QSE *B* sells 1 MW of regulation up with a MCPC of $11/MW, and QSE C buys 1 MW of energy with LMP of $29/MWh.

Once the DAM has solved, ERCOT communicates the awarded energy offers and bids, CRR offers, PTP obligation bids, and ancillary service offers to the market participants.

3 RUC

3.1 RUC Process Overview

In short, RUC ensures that there is enough resource capacity, in addition to ancillary service capacity, committed in the right locations to reliably serve the forecasted load on the ERCOT system as shown in Fig. 2.14. These processes ensure that the system is secure through the commitment and de-commitment of resources in addition to all previous commitments made by QSEs.

Specifically, the RUC process considers capacity requirements for each hour of the study period and recommends commitment of generation resources needed to reliably serve ERCOT's forecasted load, subject to all transmission constraints and resource performance characteristics. The RUC process delivers a solution that ensures enough capacity is committed in the right locations for the reliability of the ERCOT grid while minimizing costs.

Fig. 2.14 Illustration of
RUC

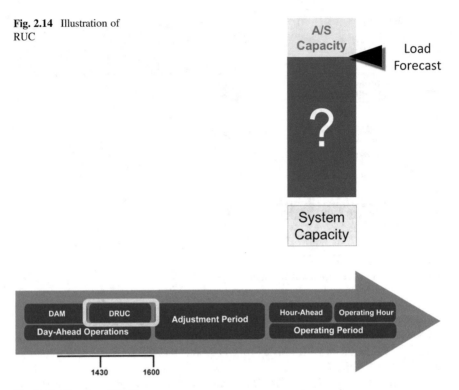

Fig. 2.15 Timeline of RUC

There are two reliability unit commitment sequences: day-ahead and hourly—see Fig. 2.15.

First, we'll look at the day-ahead reliability unit commitment, or DRUC, process. Remember that the DAM is executed at 10:00 a.m. and its results are required to be posted no later than 1:30 p.m., or 1330. DRUC is executed daily at 2:30 p.m., or 1430, which is a minimum of 1 hour after the publication of the DAM results. DRUC is completed by 4:00 p.m. or 1600 when resource commitments and de-commitments are communicated to QSEs.

DRUC is executed once a day, at 1430 of the day-ahead operation period, as shown in Fig. 2.16. The study period for the day-ahead reliability unit commitment is the next operating day. The DRUC ensures that enough capacity is committed for all 24 hours of the next operating day. The DRUC is ERCOT's first look at the reliability needs of a given operating day.

The other instance of RUC is the hourly reliability unit commitment process or HRUC as shown in Fig. 2.17. The hourly reliability unit commitment process is executed at the beginning of each operating period and is executed to fine-tune resource commitments made in the day-ahead reliability unit commitment process. Like the day-ahead reliability unit commitment process, the hourly reliability unit commitment process also ensures that enough capacity is committed in the right

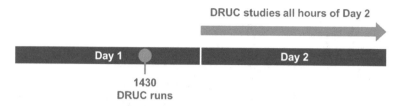

Fig. 2.16 Timeline of DRUC

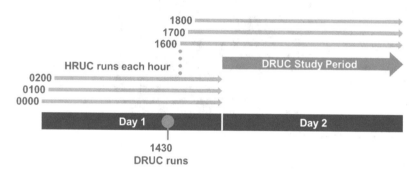

Fig. 2.17 Timeline of HRUC

locations to reliably cover the forecasted system load. However, its study period is different from DRUC.

The HRUC process has two different study periods, depending on when it is run. As we stated earlier, the HRUC process fine-tunes resource commitment decisions already made by DRUC, and its evaluation is to determine if any more commitments need to be made. Therefore, HRUC will not evaluate the need for resource commitments for a given operating hour that has not already been evaluated by DRUC. Accordingly, if the DRUC process has not been executed yet for the next operating day, then the HRUC will only evaluate the remaining hours of the current operating day.

Once DRUC has been executed, then the HRUC process will evaluate the balance of the current operating day and all the operating hours of the next operating day. As it can be seen, a given operating hour will be studied for resource commitment needs, once by DRUC and many more times by the HRUCs executed after the DRUC. The last HRUC that evaluates a given operating hour will be executed at the end of the adjustment period for that hour.

3.2 Input to RUC Process

Regardless of whether we are running the DRUC or the HRUC (or the week-ahead RUC for that matter), the process is the same. As we can see, there are many inputs to

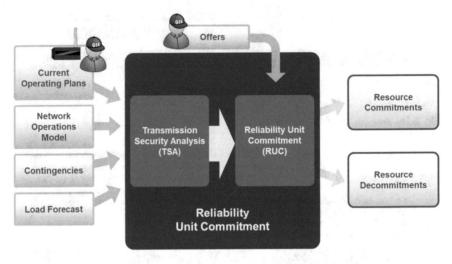

Fig. 2.18 Reliability unit commitment process

the RUC process as shown in Fig. 2.18. There are two essential parts to the process: transmission security analysis and reliability unit commitment. After the RUC process is completed, the outputs are resource commitments or resource de-commitments, if any are needed at all.

3.2.1   Current Operating Plan (COP)

The COP is an hourly plan submitted by a QSE reflecting anticipated operating conditions for each of the resources that it represents for each hour in the next seven operating days.

QSEs must keep their current operating plans updated at all times. The RUC process will look at the status fields of each QSE's current operating plans to determine if resources are available for commitment. The two possible values for the status of resources eligible for RUC commitments are "OFF" and "ONRR." "OFF" indicates that the resource is offline but available for commitment by the DAM or RUC. "ONRR" indicates that the resource is online as a synchronous condenser and available for commitment by RUC. Any other resource status means that the resource is already planned to be online and thus does not need to be committed by RUC or that the resource is not available for RUC commitment. Remember, the QSE must update the current operating plans by 1430, prior to the execution of DRUC.

3.2.2 Network Operations Model

The network operations model, maintained by ERCOT, depicts the normal conditions of the grid. It reflects the normal topology or expected connectivity and contains all of the equipment ratings. The model also defines certain remedial action schemes, automatic mitigation plans, and remedial action plans.

The RUC retrieves the transmission outage information from outage scheduler (OS) and uses the outage information to build the network topology.

The RUC retrieves dynamic ratings data from the EMS for transmission equipment when available. The dynamic rating is used in network security monitor (NSM) function. The RUC uses default static rating data from the EMS for the transmission equipment when the dynamic rating data is not available. The dynamic ratings are weather-adjusted MVA limits for each hour of the study period for all transmission lines and transformers.

3.2.3 Generic Transmission Constraints

Generic transmission constraints are the network/voltage constraints modeled as import/export energy constraints that are determined offline. Generic transmission constraints represent stability and voltage limits between areas and, in general, are more constraining than thermal limits. One example is shown in Fig. 2.19.

To have a clear understanding of generic transmission constraints, some concepts need to be introduced.

Contingency: An event that could jeopardize reliability.
Constraint: A system limit that prevents us from using the cheapest power available.
Congestion: When we must adjust power flows to keep from overloading a constraint under a specific contingency, we say that the system is experiencing congestion.

A violation of generic transmission constraints can result in the security violations, i.e., we mean an overloaded constraint under a specific contingency. Therefore, security violations will always invoke a contingency and constraint pair.

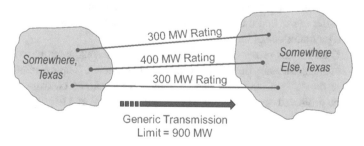

Fig. 2.19 Generic transmission constraints

It is critical to consider generic transmission constraints when running RUC process to ensure that the grid reliability and stability is preserved.

3.2.4 Load Forecast

The DAM is not working under the constraint of serving the forecasted load of the ERCOT system. The DAM is voluntary, and loads are not required to participate.

The RUC retrieves the most current hourly load forecast from the mid-term load forecast (MTLF) application from EMS for each weather zone. The MTLF predicts the hourly loads for the next 168 hours based on current weather forecast parameters within each weather zone. The ERCOT system-wide load forecast is calculated as the sum of load forecasts for all the weather zones. The accuracy of the load forecast is critical to the RUC since it is used by RUC to secure generation capacity.

The RUC also retrieves the hourly load distribution factors from EMS for each hour in the study period. Load distribution factors map out the location or distribution of forecasted load across the electrical buses in the network model. The distribution of load across a network model will determine the power flow in the modeled ERCOT system. This distribution is necessary for RUC to ensure enough generation is available in the right areas.

Load distribution factors are calculated based on:

- Distribute the load forecast to individual buses within a load zone.
- Based on historical power flows.
- Load distribution factors are published based on "cold," "mild," or "hot" proxy days.

3.2.5 Offers and Proxy Energy Offer

If additional resources need to be committed, RUC will choose the most economic resources. The source of this economic information is the three-part supply offer for any units that RUC might be considering. However, RUC does not use the entire three-part supply offer. When evaluating a resource for commitment, RUC considers the startup offer and the minimum-energy offer of a resource. After all, these two costs represent the costs to bring the resource online in response to a commitment. The energy offer curve is not used in the RUC evaluation but may be applied in the make-whole payment calculation.

The RUC calculates proxy energy offer curves for all resources based on their mitigated offer caps to substitute their original offer curves. The calculated proxy energy offer curves will then be used by the RUC market clearing engine to determine the projected energy output level of each resource and to project potential congestion patterns for each hour of the RUC.

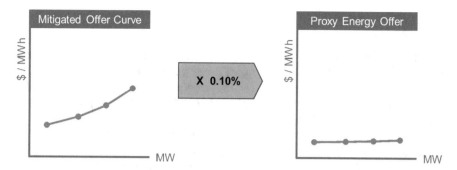

Fig. 2.20 Proxy energy offer curve, obtained by using mitigated offer cap curves for the generation resource type and multiplying it by a small constant ($< 0.10\%$) selected by ERCOT for every point between HSL and LSL

The RUC calculates the proxy energy offer curves by multiplying the mitigated offer cap by a configurable discount parameter and applies the cost for all generation output between high sustained limit and low sustained limit.

The proxy energy offer curve is calculated in a way that discounts the incremental energy cost to ensure the self-committed online resources are used to the fullest extent before additional RUC commitment as illustrated in Fig. 2.20. This approach will minimize the out-of-the-market reliability commitment which may overmitigate the competitive market price. In turn, this approach will also minimize the RUC make-whole payment resulting from the startup and minimum-energy cost.

3.3   RUC Process

The security-constrained unit commitment (SCUC) program is the core solution engineer used by RUC to determine the optimal commitment schedules. The SCUC engine is composed of two major functional components:

- Network-constrained unit commitment (NCUC)
- Transmission security analysis (TSA)

The NCUC function is used to determine projected commitment schedules that minimize the total operation costs over the RUC study period while meeting forecast demand subject to transmission constraints and resource constraints. These resource constraints represent the physical and security limits on resources.

- High sustained limit and low sustained limit
- Minimum online time
- Maximum online time
- Minimum offline time
- Maximum offline time

- Maximum daily startup
- Startup time

The objective function of NCUC is defined as the sum of startup cost, minimum-energy cost, and incremental energy cost based on three-part supply offers while substituting a proxy energy curve for the energy offer curve. The NCUC also employs the penalty factors on violation of security constraints in the objective function to ensure feasible solution.

Normally, the NCUC is executed several times iterating with TSA within SCUC solution process. The first NCUC run is referred to as the initial unit commitment (IUC). IUC has the same function as NCUC except that the network constraints are not considered. In the remaining NCUC executions, the network constraint data is prepared by TSA. The network constraint data is then passed to the NCUC as additional constraints for enforcement in the optimization process.

The function of network security analysis is to evaluate the feasibility of the generation schedule for the base case network as well as for the post-contingency network states. AC power flow results are used to evaluate the feasibility of the base case. A linearized analysis is used to evaluate the impact of the postulated contingencies. The TSA is performed for each interval of the RUC study period to provide constraints for the NCUS function. For each violated constraint, a set of shift factors are calculated with respect to generation resources. The shift factors and violated constraints are passed to the NCUC for enforcement.

As shown in Fig. 2.21, the RUC problem is solved iteratively as follows.

First, the RUC process determines initial unit commitment. It includes resources previously committed and may add commitments to meet load forecast. In this step, it does not recognize transmission constraints.

Fig. 2.21 Solving RUC process

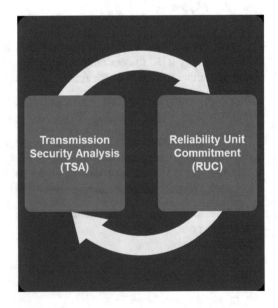

Second, the RUC process produces a dispatch solution for input to transmission security analysis using proxy energy offer curve. It allows RUC to calculate a dispatch solution to project congestion patterns for TSA. The proxy energy offer curve is derived from mitigated offer caps and used to estimate projected energy output level per generation resource and project potential congestion patterns per hour within RUC. The proxy energy offer curve can be considered as the "normalized" energy cost contribution of generation resources of similar technology and fuel, therefore making economic selection based on startup and minimum-energy cost only.

Third, the RUC process checks to see if dispatch solution is secure by testing base case and contingency cases. If TSA determines transmission constraints, it produces a set of security violations to be solved by RUC.

Fourth, the RUC determines a revised unit commitment to enforce transmission constraints.

Last, the RUC repeat process to revise resource commitments as needed to resolve security violations until solutions converge.

3.3.1 RUC Problem Formulation and Solution Algorithm

The incremental energy offer curves in RUC are piecewise linear. Hence, the RUC optimization cost functions calculated from the integration of the incremental energy offer curves become quadratic. This makes the RUC optimization problem as a mixed-integer quadratic programming (MIQP) problem. Compared to the linear MIP model, the MIQP model usually takes longer time to solve, and the performance is not desirable for the large RUC problem. To solve this issue, the following two-step solution algorithm has been proposed for the NCUC.

- Unit commitment (UC) step
- Economic dispatch (ED) step

First, the UC step uses mixed-integer programming (MIP) to solve the overall problem by treating the piecewise linear incremental cost curves as staircase constant. The number of horizontal segments used for the approximation is determined in such a way that the areas between the original curve and the staircase segment are less than a user-defined constant. The UC step determines the unit commitment schedules and binary variables including the unit commitment decision variables and the ones used for the startup cost modeling (hot, intermediate, and cold startup). The results of the binary variables from UC steps are passed to ED step as fix values.

Next, the ED step uses the quadratic programming (QP) to solve the same problem again with the original cost curve but with all the binary variables fixed at the solutions of the UC step. The ED step determines the unit dispatch schedules and market clearing prices.

RUC is executed to ensure transmission system reliability and to ensure that enough resource and ancillary service capacity is committed in the right location to reliably serve the forecasted load on the ERCOT system. Like DAM, the RUC market clearing problem is a security-constrained unit commitment that is formulated as a mixed-integer linear program (MILP) problem.

The RUC problem is formulated as follows:

Minimize:

$$\sum_{r,t} \text{StartupCost}_{r,t}$$

$$+ \sum_{r,t} \text{MineneCost}_{r,t} {}^* \text{ResProcured}_{r,t}$$

$$+ \sum_{r,\text{seg},t} \text{ProxyEnergyCurve}_{r,\text{seg},t} {}^* \text{ResSegMW}_{r,\text{seg},t}$$

$$+ \text{EnergyPenalty}^* \sum_t (\text{EnergySlackUPMW}_t + \text{EnergySlackDNMW}_t)$$

$$+ \sum_{r,\text{astype},t} \text{ClearedMWSlack}_{r,\text{astype},t} {}^* \text{ASPenalty}_{\text{astype}}$$

$$+ \sum_c \text{TranPenalty}_c {}^* \sum_t \text{TranSlackMW}_{c,t}$$

$$+ \text{WeeklyEnergyPenalty}^* \sum_r \text{SlackWeeklyEnergy}_r$$

Subject To:

1. *Power balance constraints for each interval t:*

$$\sum_{r,\text{seg}} \text{Re sSegMW}_{r,\text{seg},t} + \sum_r \text{MinCapMW}_{r,t} * \text{Re sProcured}_{r,t}$$
$$+ \text{EnergySlackUPMW}_t - \text{EnergySlackDNMW}_t$$
$$= \text{loadforecast}_t (\text{SP}_{\text{demand},t})$$

2. *Network constraints at each interval t for both base cases and contingency cases:*

$$\sum_{\text{Online Resource } j} \text{SF}_{j,c,t} {}^* \text{ResDispMW}_{j,t}$$

$$+ \sum_{r,\text{offer}} \text{SF}_{r,c,t} {}^* \left(\text{MinCapMW}_{r,t} {}^* \text{ResProcured}_{r,\text{offer},t} + \sum_{\text{seg}} \text{ResSegMW}_{r,\text{offer},\text{seg},t} \right)$$
$$- \text{TranSlackMW}_{c,t} \le \text{Limit}_{c,t}.$$
$$(\text{SP}_{c,t})$$

3. *Operating mode constraints for each resource r at each interval t:*
   $\text{ResProcured}_{r,\,t} = 1$ if mode is "R."
   $\text{ResProcured}_{r,\,t} = 0$ if mode is "U."

4. *Resource constraints for each resource r:*
   For $t \geq 2$:

$$\mathrm{Re\,sProcured}_{r,t} - \mathrm{Re\,sProcured}_{r,t-1} = \mathrm{Re\,sStartup}_{r,t} - \mathrm{Re\,sShutdown}_{r,t}$$

   For $t = 1$:

$$\mathrm{Re\,sProcured}_{r,t=1} - \mathrm{Re\,sInitStatus}_r = \mathrm{Re\,sStartup}_{r,t=1} - \mathrm{Re\,sShutdown}_{r,t=1}$$

   For $t \geq 1$:

$$\mathrm{Re\,sStartup}_{r,t} + \mathrm{Re\,sShutdown}_{r,t} \leq 1$$

5. *Resource minimum online/offline time constraints for each resource r:*
   For $t \leq T - 1$:

$$[\min(T - t, (\mathrm{mnuptm}_r - 1))]\,\mathrm{Re\,sStartup}_{r,t}$$
$$\leq \quad \mathrm{Re\,sProcured}_{r,t+1} + \mathrm{Re\,sProcured}_{r,t+2} + \ldots$$
$$+ \mathrm{Re\,sProcured}_{r,t+\min[T - t\mathrm{mnuptm}_r - 1]}$$

   For $t \leq T - 1$:

$$[\min(T - t, \mathrm{mndntm}_r - 1)]\,\mathrm{Re\,sShutdown}_{r,t} + \mathrm{Re\,sProcured}_{r,t+1}$$
$$+ \mathrm{Re\,sProcured}_{r,t+2} + \ldots \quad + \mathrm{Re\,sProcured}_{r,t+\min[T - t,\mathrm{mndntm}_r - 1]}$$
$$\leq \min(T - t, \mathrm{mndntm}_r - 1)$$

6. *Resource initial condition constraint for resource r:*

$$\mathrm{Re\,sProcured}_{r,t} = 1, \quad t = 1, \ldots, \min[T, \mathrm{mnuptm}_r - \mathrm{Re\,sHrinitUp}_r]$$
$$\text{if} \quad \mathrm{Re\,sInitStatus} = 1$$

$$\mathrm{Re\,sProcured}_{r,t} = 0, \quad t = 1, \ldots, \min[T - 1, \mathrm{mndntm}_r - \mathrm{Re\,sHrinitDn}_r]$$
$$\text{if} \quad \mathrm{Re\,sInitStataus} = 0$$

$$\sum\nolimits_{t\,\mathrm{incurrentday}} \mathrm{Re\,sStartup}_{r,t} + \mathrm{Re\,sNumstartupinit}_r \leq \mathrm{DailyStartuplimit}_r$$

7. *Resource maximum daily startup constraint for resource r:*

$$\sum\nolimits_{t\,\mathrm{innextday}} \mathrm{Re\,sStartup}_{r,t} \leq \mathrm{DailyStartuplimit}_r$$

8. *Resource maximum runtime constraint for resource r:*
   For $t = 1$:

$$\mathrm{Re\,sHrsUp}_{r,t=1} \leq \mathrm{Re\,sHrinitUp}_r + 1$$

$$\mathrm{ResHrsUp}_{r,t=1} \geq \mathrm{ResHrinitUp}_r + 1 - (1 - \mathrm{ResProcured}_{r,t=1})$$
$$* (\mathrm{T} + \mathrm{ResHrinitUp}_r + 1)$$

$$\mathrm{Re\,sHrsUp}_{r,t=1} \leq \mathrm{Re\,sProcured}_{r,t=1} * (T + \mathrm{Re\,sHrinitUp}_r + 1)$$

For $t \geq 2$:

$$\mathrm{Re\,sHrsUp}_{r,t} \leq \mathrm{Re\,sHrsUp}_{r,t-1} + 1$$

$$\mathrm{Re\,sHrsUp}_{r,t} \geq \mathrm{Re\,sHrsUp}_{r,t-1} + 1 - (1 - \mathrm{Re\,sProcured}_{r,t})$$
$$* (T + \mathrm{Re\,sHrinitUp}_r + 1)$$

$$\mathrm{Re\,sHrsUp}_{r,t} \leq \mathrm{Re\,sProcured}_{r,t} * (T + \mathrm{Re\,sHrinitUp}_r + 1)$$

For $t \geq 1$:

$$\mathrm{Re\,sHrsUp}_{r,t} \leq \mathrm{Re\,sMaxRuntime}_r$$

9. *Downtime-dependent startup cost for resource r:*
   For $t = 1$:

$$\mathrm{Re\,sHrsDn}_{t=1} = \mathrm{Re\,sHrinitDn}_r$$

For $t \geq 2$:

$$\mathrm{Re\,sHrsDn}_t \leq \mathrm{Re\,sHrsDn}_{r,t-1} + 1$$

For $t \geq 2$:

$$\mathrm{Re\,sHrsDn}_t \leq (T + 1 + \mathrm{Re\,sHrinitDn}_r) * (1 - \mathrm{Re\,sProcured}_{r,t-1})$$

For $t \geq 2$:

$$\mathrm{Re\,sHrsDn}_t \geq \mathrm{Re\,sHrsDn}_{r,t-1} + 1 - (T + 1 + \mathrm{Re\,sHrinitDn}_r)$$
$$* \mathrm{Re\,sProcured}_{r,t-1}$$

$$\mathrm{StartupCost}_{r,t} \geq \mathrm{Re\,sStartupcostOffer}_{r,t}(\mathrm{Re\,sHrsDn}_{r,t}) - \text{max startupcost}$$
$$* (2 - \mathrm{Re\,sStartup}_{r,t} - \mathrm{StartupcostEligibility}_{r,t}) \quad \mathrm{StartupCost}_{r,t}$$
$$\leq \text{max startupcost} * \mathrm{Re\,sStartup}_{r,t}$$

$$\mathrm{StartupCost}_{r,t} \leq \text{max startupcost} * \mathrm{StartupcostEligibility}_{r,t}$$

10. *Operating limits for each resource r at interval t:*

$$0 \leq \mathrm{Re\,sSegMW}_{r,\mathrm{seg},t} \leq \mathrm{ProxyEnergyCurveSegSize}_{r,\mathrm{seg},t} * \mathrm{Re\,sProcured}_{r,t}$$

$$0 \leq \sum_{seg} \mathrm{Re\,sSeg}_{r,\mathrm{seg},t} \leq (\mathrm{MaxCapMW}_{r,t} - \mathrm{MinCapMW}_{r,t}) * \mathrm{Re\,sProcured}_{r,t}$$

11. *Startup cost eligibility for resource r at interval t:*
    Assume that resource's self-commitment starts up the resource at interval t'.
    For each interval t before t',
    For $t < t'$:

$$\left[(t' - t) - \mathrm{Re\,sStartup}_{r,t} - \sum_{k=t+1}^{k=t'-1} \mathrm{Re\,sProcured}_{r,k} \right] - \mathrm{startupcostEligibility}_{r,t}$$
$$* T \leq 0$$

$$\left[(t' - t) - \mathrm{Re\,sStartup}_{r,t} - \sum_{k=t+1}^{k=t'-1} \mathrm{Re\,sProcured}_{r,k} \right] \geq \mathrm{startupcostEligibility}_{r,t}$$

For $t \geq t'$:

$$\mathrm{startupcostEligibility}_{r,t} = 1$$

For resource with no self-commitment, for all intervals *t:*

$$\mathrm{startupcostEligibility}_{r,t} = 1$$

12. *Protection of resource capacity providing ancillary service for resource r at interval t:*

$$(\mathrm{LASL}_{r,t} - \mathrm{ClearedMWSlack}_{r,\mathrm{DRS},t}) - (1 - \mathrm{ResProcured}_{r,t}) * \mathrm{MaxCapMW}_{r,t}$$
$$\leq \mathrm{MinCapMW}_{r,t}{}^{*}\mathrm{ResProcured}_{r,t} + \sum_{seg} \mathrm{ResSegMW}_{r,t}$$

$$\mathrm{HASL}_{r,t} + \sum_{\mathrm{astype}\,=\,\mathrm{URS,\,RRS,\,NSRS}} \mathrm{ClearedMWSlack}_{r,\mathrm{astype},\,t}$$
$$\leq$$
$$+ \left(1 - \mathrm{ResProcured}_{r,t}{}^{*}\mathrm{MaxCapMW}_{r,t} \right)$$

$$\sum_{\mathrm{astype}\,=\,\mathrm{DRS,URS,RRS,NSRS}} \mathrm{ClearedMWSlack}_{r,\mathrm{astype},t} \leq \mathrm{Re\,sProcured}_{r,t}$$
$$* \mathrm{MaxCapMW}_{r,t}$$

13. *Some weekly constraint for resource r in WRUC:*

$$\sum_t \text{MinCapMW}_{r,t} * \text{Re sProcured}_{r,t} + \sum_{\text{seg},t} \text{Re sOfferSeg}_{r,seg,t}$$

$$- \text{SlackWeekenergy}_r \leq \text{WeeklyEnergylimit}_r$$

$$\sum_t \text{Re sStartup}_{r,t} \leq \text{WeeklyStartuplimit}_r$$

14. *Ignore temporal constraints violated by self-commitment:*
   Do not apply minimum up−/downtime in 5), maximum runtime in 8), and daily/weekly startup and weekly energy constraints in 7) and 13) to the self-commitment period. As a consequence, the initial condition constraint 6) is also ignored.

15. *Combined cycle resource configuration constraints:*
   All the resource-level constraints defined above for resource *r* remain applicable for any possible configuration for combined cycle generation resource. In other words, in case of combined cycle resource *r* represents configuration *r* of the combined cycle resource.
   The additional inter-configuration constraints for combined cycle resource **ccunit** with multiple configurations *mod* at *t* include:

$$\text{Re sProcured}_{\text{ccunit},t} = \sum_{\text{mod}} \text{Re sProcured}_{\text{ccunit,mod},t} \leq 1,$$

$$\text{Re sMinRuntime Re s}_{\text{ccunit,mod},t} \text{ Re s}_{\text{ccunit,mod ccunit,mod}}$$

   ResProcured$_{\text{ccunit}, t}$ is subjected to combined cycle plant minimum downtime and minimum runtime constraints.
   The inter-configuration transition is feasible between certain configurations and between certain configurations to the off state.

16. *Split generation resource constraints:*
   RUC commits or de-commits all individual split generation resources in a generation together. For example, resource *r1* and resource *r2* are split generation resources of generation resource; then

$$\text{Re sProcured}_{r1,t} = \text{Re sProcured}_{r2,t}$$

17. *Startup time (lead time) constraints:*
   The time required to start up a unit varies depending on how long the unit has been offline. The startup time of resources effects their eligibility to be committed as follows:

   If the resource is offline based on the EMS data, then the resource is made unavailable for commitment from the current time to the min (current time + start time, first hour resource scheduled on in the COP).

This constraint is handled as part of NCUC input data pre-processing. The operating modes for the study intervals that resources are not eligible to be committed will be set to unavailable ("U").

The RUC problem input parameters are:

- T is the number of study intervals.
- $Loadforecast_t$ is the ERCOT total load forecast at interval t.
- $MaxCapMW_{r,t}$ and $MinCapMW_{r,t}$ are the max and min operating limits of resource r at interval t.
- $HASL_{r,t}$ and $LASL_{r,t}$ are the high and low ancillary service limits of resource r at interval t.
- $mnuptm_{r,t}$ and $mndntm_{r,t}$ are the minimum uptime and minimum downtime of resource r.
- $ResInitStatus_r$ is the initial status of resource r at the start of the study period, 1, online and 0, offline.
- $ResMaxRuntime_r$ is the maximum runtime of resource r.
- For combined cycle resource, $ResMaxRuntime_{ccunit,\ mod}$ and $ResMinRuntime_{ccunit,mod}$ are the maximum and minimum runtime of configuration mod of combined cycle resource ccunit.
- $ResHrinitUp_r$ is the number of intervals that resource r has been online at the start of the study period. $ResHrinitUp_r$ is 0 if resource r is offline at the start of the study period.
- $ResHrinitDn_r$ is the number of intervals that resource r has been offline at the start of the study period; $ResHrinitDn_r$ is 0 if resource r is online at the start of the study period.
- $ResNumstartupinit_r$ is the number of daily startups of resource r at the start of the study period.
- $WeeklyStartuplimit_r$ is the maximum weekly startup limit of resource r.
- $WeeklyEnergylimit_r$ is the weekly energy limit of resource r.
- $ProxyEnergyCurve_{r,seg,t}$ is the price of MW segment seg of the proxy energy curve of resource r at interval t.
- $ProxyEnergyCurveSegSize_{r,seg,t}$ is the MW size of segment seg of the proxy energy curve of resource r at interval t.
- $ResStartupcostOffer_{r,t}$ is the startup cost offer of resource r with three-part offer at interval t.
- $maxstartupcost$ is the maximum startup cost of the startup cost offers at all intervals of all resource r.
- $MineneCost_{r,t}$ is the minimum-energy offer of resource r with three-part offer at interval t.
- $Limit_{c,t}$ is the NSM calculated limit of transmission constraint c at interval t. This limit represents the aggregated effect of the latest control variables on the constraint and the violation on the constraint.
- $SF_{r,c,t}$ is the shift factor of resource r associated with transmission constraint c at interval t.
- $EnergyPenalty$ is the penalty cost of violating the power balance constraint.

- ASPenalty$_{astype}$ is the penalty cost of violating the AS protection for AS type astype.
- TranPenaltyCost$_c$ is the penalty cost of violating the transmission constraint c.
- WeeklyEnergyPenalty is the penalty cost of violating the resource weekly energy limit.

The RUC problem decision variables are:

- *Integer Variables:*

  - ResProcured$_{r,\,t}$ is 1 if resource r is online for procurement at interval t and 0 offline.
  - For combined cycle resource, ResProcured$_{ccunit,mod,t}$ is 1 if configuration mod of combined cycle resource ccunit is online for procurement at interval t and 0 offline.
  - For combined cycle resource, ResProcured$_{ccunit,t}$ is 1 if any of the multiple configurations of combined cycle resource ccunit is online for procurement at interval t and 0 otherwise.
  - ResStartup$_{r,t}$ is 1 if resource r is started up at interval t and 0 otherwise.
  - ResShutdown$_{r,t}$ is 1 if resource r is shut down at interval t and 0 otherwise.
  - ResHrsUp$_{r,t}$ is the number of intervals that resource r has been online at the end of interval t.
  - ResHrsDn$_{r,\,t}$ is the number of intervals that resource r has been offline at the start of interval t.
  - StartupcostEligibility$_{r,t}$ is 1 if resource r's startup cost is eligible at interval t and 0 otherwise.

- *Continuous Variables:*

  - ResSegMW$_{r,seg,t}$ is the scheduled MW of segment seq of proxy energy offer of resource r at interval t.
  - StartupCost$_{r,t}$ is the startup cost for resource r at the start of interval t.

- *Dual Variables (Shadow Prices):*

  - SP$_{demand,\,t}$ is the dual variable associated with the power balance constraint at interval t.
  - SP$_{c,\,t}$ is dual variable associated with the transmission constraint c at interval t.

- *Non-negative Slack Variables:*

  - EnergySlackUPMW$_t$ and EnergySlackDNMW$_t$ are the slack variables for the power balance constraint at interval t.
  - ClearedMWSlack$_{r,astype,t}$ is the slack variable for protecting resource r capacity for AS type astype at interval t.
  - TranSlackMW$_{c,t}$ is the slack variable for the transmission constraint c at interval t.
  - SlackWeeklyEnergy$_r$ is the slack variable for the weekly energy constraint of resource r.

3.4   Energy Offer Curve for RUC-Committed Resource

If a resource is RUC-committed, ERCOT will communicate the commitment to the QSE. The communication will include the start interval and duration, for which the resource is required to be at least at the low sustained limit (LSL). Although most of this communication will be through an automated messaging system, in some cases, ERCOT may communicate dispatch instructions verbally. Whether the communication is electronic or verbal, the QSE must acknowledge receipt of the instruction and update its COP to reflect the new status.

The QSE may update their energy offer curve (EOC) prior to the RUC commitment period while EOC for all energy above LSL subject to offer floor. ERCOT will adjust EOC if the QSE does not adjust the EOC. This only applies only to hours in the RUC commitment period—see Fig. 2.22.

4   Real-Time SCED

The goal of real-time operations is to manage the reliability of the ERCOT transmission system while operating at the least cost as shown in Fig. 2.23. In order to manage reliability, the real-time process must dispatch resources to balance generation with demand while operating the transmission system within established limits. To minimize costs, the real-time process evaluates the economics of each online resource and determines the least cost dispatch to meet reliability needs.

In managing the system reliability, there are several constraints that must be considered. ERCOT's first goal is to achieve power balance, in which generation is dispatched to match demand. Transmission constraints identify the flow limits of the transmission system. These limits can be thermal in nature, but they can also be

Fig. 2.22 EOC subject to an offer floor

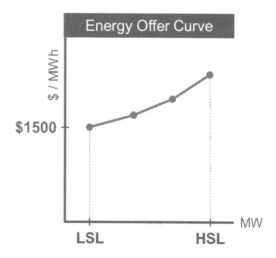

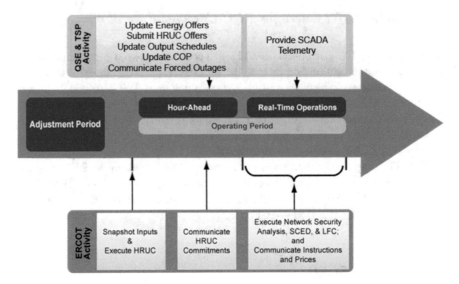

Fig. 2.23 The operating period activity timeline

related to system stability. Resource constraints influence how ERCOT dispatches resources to serve the demand and to manage transmission congestion. While ERCOT assures that the transmission system is operated within its limits, it must also assure that both generation and load resources are dispatched within their operational capabilities.

There are two contributing factors that influence these constraints. Changing weather conditions can affect the system load, but the weather also affects the capacity of the transmission system and the performance of resources. Severe weather can even affect the availability of transmission elements and resources needed to serve the load. Planned and unplanned outages reduce the amount of transmission capacity and resource capacity that are available to serve load.

The real-time process can be divided into two categories: energy dispatch and load frequency control. In the energy dispatch process, ERCOT will achieve power balance and manage congestion while simultaneously achieving a least cost dispatch of generation resources. Load frequency control is used to maintain power balance in between energy dispatches.

4.1   SCED: Energy Dispatch

4.1.1   Information Flows in SCED

Figure 2.24 shows how information flows in and out of these SCED processes.

First, real-time system data is provided to the real-time operations process through telemetry. QSEs communicate resource limits, statuses, and output levels,

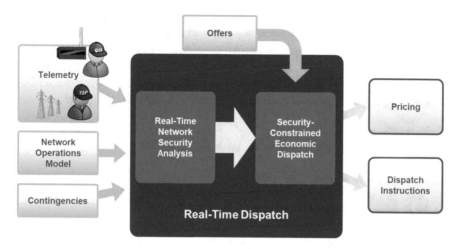

Fig. 2.24 SCED overview

while transmission service providers communicate current flows on transmission lines and voltages at various locations throughout the system.

Telemetered data for each resource provided in real time by QSEs to ERCOT includes information such as the real and reactive power output of generation resources. The high and low sustained limits, initially communicated through the COP, are now provided through telemetry. QSEs also provide corresponding information for load resources.

The actual status of a resource during the operating hour is known through the telemetered breaker and switch statuses. Telemetry also indicates how much of each ancillary service is available on each resource. The ancillary service resource responsibility, initially communicated through the current operating plan, indicates the amount of an ancillary service that is reserved on a specific resource. The ancillary service schedule indicates how much of that reserved capacity is currently available to be deployed. Finally, real-time telemetry will indicate the current configuration of combined cycle resources.

Second, ERCOT systems provide other important information such as network topology and equipment ratings. Most of this information is contained in the network operations model, although dynamic ratings may be provided through real-time telemetry.

Finally, the real-time operations process receives a list of contingencies to study. These are "what-if" scenarios that ERCOT wants the real-time operations processes to study to insure reliable operation of the grid after system events such as transmission or resource outages that are known to have a significant impact on the grid.

A QSE may de-commit a resource for intervals that were not RUC-committed:

- A QSE may de-commit a quick start generation resource (QSGR) without request.
- For all other resources, the QSE must verbally request permission from ERCOT.

In some cases, a resource is forced offline. When the forced outage occurs, ERCOT is automatically notified of the event through real-time telemetry. In the case of a forced outage, impending forced outage, or de-rating of a resource, the QSE must inform ERCOT of the nature of the incident. Additionally, the QSE should indicate when the resource is expected to be online again, or, with an impending forced outage, the QSE should indicate the time the QSE expects the outage to occur.

4.1.2 Real-Time Network Security Analysis

Rea-time network security analysis is performed within two steps.

First, the state estimator takes all the real-time telemetry combined with the network operations model.

Ultimately, the state estimator provides a consistent picture of breaker statuses, line flows, and bus voltages for us to study for security violations.

Next, network security analysis performs a contingency analysis of the ERCOT system. It takes each contingency from our standard contingency list and inflicts them on the system, one at a time. In each contingency case, it predicts the post-contingency loading of all transmission elements in the system. If any element is loaded past its emergency rating, there is a security violation. This contingency analysis is the basis of maintaining an "$N - 1$" level of security, i.e., the loss of a single element doesn't immediately overload other elements.

If the contingency analysis discovers a security violation, there is an $N - 1$ constraint. Before sending the constraint over to SCED to solve, operators have a manual checkpoint.

For SCED to resolve a constraint, the constraint must be activated in SCED. All constraints must be verified as accurate and appropriate before being activated in SCED. Any constraints that are commonly seen are activated immediately. Novel or infrequent constraints are reviewed before they are activated in SCED. In such a case, SCED will be allowed to run on its normal schedule. The constraint simply would not be activated in SCED until a subsequent SCED run.

Once the real-time network security analysis is complete, it passes SCED two types of information: constraints and shift factors. The constraints are simply the transmission elements whose loading needs to be reduced. Shift factors tell SCED the impact that each generator has on each constraint.

4.1.3 Resource Limit Calculator (RLC)

The resource limit calculator requires each resource's high sustainable and low sustainable limits, as well as its current operating point. This information is telemetered by the QSE every few seconds. The HSL and LSL describe the current maximum and minimum sustained energy production capability of the resource. They set the upper and lower boundaries within which the calculator will determine the rest of its results.

Table 2.3 Constraints in SCED	SCED constraint	Input to SCED
	Power balance constraint	Generation to be dispatched
	Transmission constraints	Constraints and shift factors
	Resource constraints	Resource limit calculator

The commitment of a resource's capacity to providing ancillary services restricts its ability to provide energy for normal energy dispatch. When calculating dispatch limits, the RLC takes into account a resource's ancillary services schedule, current telemetry, and high and low sustainable limits and calculates a new set of limits to reflect this. These limits define the range of base points to which a resource can be deployed by SCED while maintaining its ability to provide ancillary services. These limits are known as the high ancillary service limit (the HASL) and the low ancillary service limit (the LASL).

Once a resource's SCED-ramp rates are defined, RLC multiplies each by the length of the SCED interval (normally, 5 minutes). It then adds the SCED-Up ramp rate to the resource's current telemetry to render a high dispatch limit (called the HDL). Similarly, SCED subtracts the SCED-Down ramp rate from the resource's current telemetry to determine the low dispatch limit (called the LDL). These limits denote the region within which SCED may create a base point for dispatching the resource.

Finally, SCED need the total MW quantity of generation that needs to be dispatched. To calculate the generation to be dispatched (GTBD), it begins with the current total output from all generation resources in ERCOT and then adds or subtracts from this amount to accommodate our projected load change over the next 5 minutes. Table 2.3 shows the constraints used in SCED.

4.1.4 How SCED Functions

When we refer to the balance of reliability and economics, SCED is the instrument to achieve this balance. On the one side of the scale, SCED receives the system constraints that must be respected to meet the reliability needs of the ERCOT grid from the network security analysis process. To meet those reliability needs, SCED will have to dispatch resources on the system in some way. On the other side of the scale, SCED draws on the energy offer curves for all online resources to find the least cost dispatch of these resources. What would happen if we did not achieve this balance between reliability and economics?

The balance between reliability and economics means we are using the full available capacity of the transmission grid, using the least cost generation given the constraints of the system, and serving load at the least overall cost. This balance is the goal of the SCED process. To accomplish this goal, the SCED process evaluates energy offer curves to produce a least cost dispatch of online generation resources while respecting transmission and generation constraints.

The objective of the SCED is to minimize the total energy offer costs over the study period for online generation resources (GR), demand resources (DR), and storage resources (SR), minus the response costs for aggregate retail resources (ARR):

$$
\underset{P_r^t}{\text{Minimize}}
\begin{cases}
\sum_{t\in T}\sum_{r\in\text{GR}\cup\text{DR}\cup\text{SR}}\text{EOC}_r^t\left(P_r^t\right)-\sum_{t\in T}\sum_{r\in\text{ARR}}\text{PRC}_r^t\left(P_r^t\right)+ \\
\sum_{t\in T}C_{\text{over}}^{\text{pen}}\left(P_{\text{over}}^t\right) \\
+\sum_{t\in T}C_{\text{under}}^{\text{pen}}\left(P_{\text{under}}^t\right)+ \\
\sum_{t\in T}\sum_{l\in N}\text{SP}_l^{\max_{l;\text{viol}}^t}
\end{cases}
$$

The penalty functions for over- or under-generation and the maximal shadow prices for transmission constraints are represented explicitly in the optimization objective. The dispatching problem is also subject to the following constraints.

1. System Power Balance

   The sum of resource base points shall be equal to system load at each time interval:

$$
\sum_{r\in\text{GR}\cup\text{DR}\cup\text{SR}}P_r^t-\sum_{r\in\text{ARR}}P_r^t+P_{\text{under}}^t-P_{\text{over}}^t=LF_{\text{sys}}^t+\sum_{r\in\text{DR}}\overrightarrow{P}_r^t
$$
$$
-\sum_{r\in\text{ARR}}\overrightarrow{P}_r^t;t\in T
$$

2. Transmission Constraints

   The RTD application shall consider a set of base case, contingency, and generic constraints. The transmission constraints can be expressed in the linearized form of line power flows for each time interval:

$$
-F\max_l^{l\in N \text{ and } t\in T}\sum_{r\in\text{GR}\cup\text{DR}\cup\text{SR}}P_r^t\cdot\text{SF}_r^{l^t}\sum_{r\in\text{ARR}}P_r^t\cdot\text{SF}_{r;l;\text{viol}\max}^{l^t}
$$

3. Resource Ramping Limits

   The resource ramp rates are stepwise functions of resource power output. In general, a resource can ramp over several segments within ramping time, i.e., the resource ramping limits are piecewise linear functions of power output for a given ramping time. In this case, the upward and downward resource ramping limits can be specified as the following constraints:

$$
-\text{RL}_r^{\text{dn}}\left(P_r^t,\tau_{\text{ramp}}\right)\le P_r^{t+1}-P_r^t\le\text{RL}_r^{\text{up}}\left(P_r^t,\tau_{\text{ramp}}\right);r\in\text{GR}\cup\text{DR}\cup\text{SR} \text{ and } t\in T
$$

The resource ramping limits represent the general form of piecewise linear functions, i.e., the ramping limits determine non-convex operating set. The segment selections represent discrete decisions that must be modeled by binary variables.

4. Multi-interval Block Constraints

The energy offers can be submitted as multi-interval blocks. Only variable multi-interval block offers shall be considered by the RTD application. In this case, the base points are the same across all block time intervals, i.e.,

$$P_r^t = P_r^{t+1} = \cdots = P_r^{t+T_r^{\text{block}}} ; r \in \text{GR} \cup \text{DR} \cup \text{SR and } t \in T$$

5. Resource Dispatch Limits

Each resource shall be dispatched within its operating limits with excluded ancillary service awards, i.e.,

$$\text{LASL}_r^t \leq P_r^t \leq \text{HASL}_r^t ; r \in \text{GR} \cup \text{DR} \cup \text{SR and } t \in T$$

$$\text{HASL}_r^t = P_{\text{HSL},r}^t - P_{\text{Reg-up},r}^t - P_{\text{RRS},r}^t - P_{\text{NSRS},r}^t$$

$$\text{LASL}_r^t = P_{\text{LSL},r}^t + P_{\text{Reg-down},r}^t$$

where:

EOC_r^t is the energy offer cost for resource r at time interval t

$\text{PRC}_r^t(\cdot)$ is the price-responsive cost for aggregated retail resource r at time interval t

P_r^t is the base point for resource t at time interval t

P_{over}^t is the system over-generation at time interval t

P_{under}^t is the system under-generation at time interval t

$C_{\text{over}}^{\text{pen}}$ is the penalty function for over-generation

$C_{\text{under}}^{\text{pen}}$ is the penalty function for under-generation

SP_l^{max} is the maximal shadow price for transmission constraint

$P_{l;\text{viol}}^t$ is the violation of transmission constraint at time interval t

P_r^t is the response projection of resource r at time interval t

LF_{sys}^t is the system load forecast at time interval t

SF_r^l is the shift factor for resource r and transmission constraint

F_{max}^l is the maximal limit for transmission constraint

RL_r^{dn} is the downward ramping limit for resource r

RL_r^{up} is the upward ramping limit for resource r

τ_{ramp} is the ramping time

LASL_r^t is the low ancillary service limit for resource r at time interval t

HASL_r^t is the high ancillary service limit for resource r at time interval t

$P_{\text{HSL},r}^t$ is the high sustained limit for resource r at time interval t

$P_{\text{LSL},r}^t$ is the low sustainable limit for resource r at time interval t

$P_{\text{Reg-up},r}^t$ is the regulation up ancillary service responsibility for resource r at time interval t

$P^t_{\mathrm{RRS},r}$ is the responsive reserve responsibility for resource r at time interval t

$P^t_{\mathrm{NSRS},r}$ is the non-spinning service responsibility for resource r at time interval t

$P^t_{\mathrm{Reg-down},r}$ is the regulation down ancillary service responsibility for resource r
at time interval t

T^{block}_r is the multi-interval block for resource r

$r \in$ GR is the set of generation resources

$r \in$ DR is the set of demand resources

$r \in$ SR is the set of storage resources

$r \in$ ARR is the set of aggregate retail resources

$l \in N$ is the set of transmission constraints

$t \in T$ is the study period

Δt is the length of time interval

4.1.4.1 The Texas Two-Step

Each time SCED runs, it actually runs in two cycles, or steps, as shown in Fig. 2.25. This process has been also referred to as the "Texas two-step." The purpose of the "Texas two-step" is to ensure competition while reducing market power.

At the same time, however, SCED needs to allow high prices under the right circumstances. First, we may be in a situation where all generation available for dispatch is expensive. We may also have areas where expensive generation is needed to resolve constraints. Lastly, we may be in period of scarcity, where we are running low on dispatchable capacity. The laws of supply and demand imply that under scarcity conditions, high prices shall be allowed to occur.

In order for this process to ensure competition, transmission constraints must first be classified as competitive or non-competitive. For a starting point, at the beginning

Fig. 2.25 Texas two-step

Fig. 2.26 Illustration of price mitigation in Texas two-step

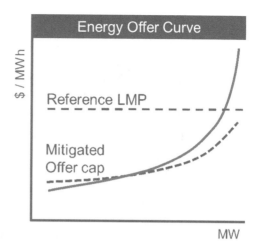

of the nodal market, the competitive constraints will be the same as the commercially significant constraints and their closely related elements on the last day of the zonal system. All other constraints will be considered non-competitive constraints.

Step 1 uses the energy offer curves for all online generators, but observes only the competitive constraints. The result is a set of reference LMPs that indicate how energy would be priced across the system if the only constraints were the competitive constraints.

Step 2, then, observes the limits of all constraints. However, SCED now invokes a price mitigation. The energy offer curve for each online generation resource capped at greater of reference LMP or mitigated offer cap—see Fig. 2.26.

This mitigated offer cap is calculated using the percentage of fuel index price and fuel oil price specified in each resource's energy offer curve and a generic or verifiable heat rate.

The result is that prices cannot be driven higher than the reference LMPs unless the verifiable costs of the resource are higher than the referenced LMPs. In either case, SCED employs price mitigation in real-time operations to mitigate the price impacts of resolving non-competitive constraints.

4.1.4.2   SCED Schedule and Output

At a minimum, SCED is executed every 5 minutes. In fact, SCED is set up on a fixed schedule to run every 5 minutes. However, it may be triggered anytime in between scheduled SCED runs, either by an operator or by other ERCOT systems—see Fig. 2.27. It is important to note that additional SCED runs will not alter the timing of the scheduled SCED runs.

As we indicated earlier, SCED produces locational marginal prices and base points. A locational marginal price (LMP) is the marginal cost of serving the next increment of load at an electrical Bus. SCED produces LMPs for each electrical bus

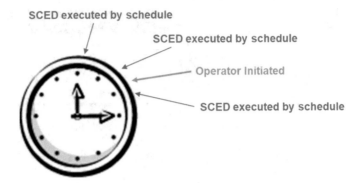

Fig. 2.27 SCED timeline

in the system (>8000). These prices are then used to calculate the settlement point prices for each resource node, load zone, and hub.

When ERCOT dispatches resource-specific base points, there is specific information that must be identified in the dispatch. ERCOT must identify the resource by name, include the desired MW output level, and indicate the time when the dispatch instruction was issued.

4.2 Load Frequency Control (LFC)

Load frequency control's sole task is to maintain system frequency. Unlike the real-time network security analysis and SCED, there is no consideration of cost optimization or other economic factors. ERCOT executes LFC every 4 seconds to restore system frequency to the scheduled frequency. The scheduled or desired system frequency is 60 cycles per second or 60 hertz. This is done by providing a control signal to each QSE that represents resources providing regulation and responsive reserve services.

Regulation service is an ancillary service that provides capacity that can respond to signals from ERCOT within seconds to respond to changes in system frequency. There are two different types of regulation services: regulation up and regulation down, commonly referred to as reg-up and reg-down.

Regulation reserve service is still deployed on a portfolio basis. ERCOT sends a portfolio-wide control signal to each QSE providing regulation. These SCADA signals go out over ICCP every 4 seconds. The QSE responds by distributing the deployment over its resources and providing ERCOT with a participation factor for each resource providing regulation. Real-time telemetry from the QSE to ERCOT includes ancillary service resource responsibilities which notify ERCOT of how much capacity has been reserved for a given ancillary service from each resource.

5 Conclusions

An efficient market could accommodate a high penetration of renewable resources while maintaining the grid reliability. A good knowledge of how such a market works is critical to understand the pitfalls and potentials of the electricity market to support the renewable integration. This chapter describes in depth the fundamental principles of such a wholesale market operated at ERCOT. ERCOT's market consists of day-ahead market, reliability unit commitment, and real-time SCED. Since 2020, ERCOT market has transitioned from a zonal-based market to a nodal market, which has greatly improved the market efficiency.

References

1. Pengwei Du, Ross Baldick, Aidan Tuohy, Integration of Large Scale Renewable Energy into Bulk Power System: From Planning to Operation, Springer, 2017
2. Pengwei Du, Ning Lu, Haiwang Zhong, Demand Response in Smart Grid, Springer, 2019
3. ERCOT, Nodal Protocol, 2022, https://www.ercot.com/mktrules/nprotocols
4. Wholesale Markets 101, ERCOT, 2022, https://www.ercot.com/services/training/courses/details?name=Wholesale%20Markets%20101%20-%20WBT

Chapter 3
Market Designs to Integrate Renewable Resources

1 Overview

The Electric Reliability Council of Texas (ERCOT) is the independent system operator that operates the electric grid and manages the deregulated wholesale electricity market for the ERCOT region. The ERCOT region covers about 75 percent of area in Texas. In 2021, ERCOT has a total of 84,404 MW of generation capacity installed. The all-time peak demand is 74,820 MW recorded on August 12, 2019. In 2001, ERCOT operates both its wholesale and retail electricity market to competition based on a nodal market structure.

Since December of 2010, ERCOT has moved from a zonal market to a locational marginal pricing (LMP)-based nodal market in which each settlement point within the ERCOT interconnection has its own prices and all transmission constraints modeled. The new nodal market should significantly increase market efficiency, better manage local congestion by issuing resource-specific instructions, and, more importantly, better integrate the rapid increase in generation capacity from Intermittent Renewable Resources (IRRs).

In the nodal market, ERCOT can better trace the individual IRR outputs since it has more accurate IRR generation forecast using weather data and the latest modeling technologies. In addition, a nodal-based market structure allows ERCOT to perform better transmission planning to integrate IRRs.

Similar to other independent system operators (ISOs) in the United States, daily wind variation in ERCOT is negatively correlated with total system load as shown in Fig. 3.1. The load tends to peak during the daytime when people are awake and using the most power for lighting, heating, and cooling and when businesses and factories are in operation. Wind generation, however, tends to peak at night. Figure 3.1 shows ERCOT hourly average system total wind generation pattern versus system total load pattern in August 2011. The average hourly wind generation may be close to minimum at the time of the daily system peak load. With very limited pumped

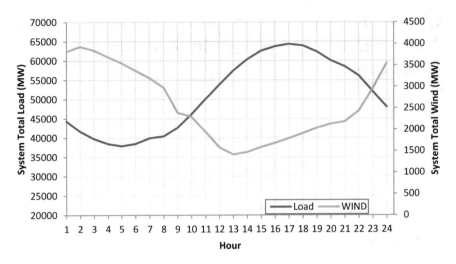

Fig. 3.1 Correlation between system load and total wind output

storage capability, ERCOT needs to manage the increasing penetration of WGRs using advanced forecasting tools and enhanced operating procedures.

ERCOT has been experiencing several operational issues due to the growth of IRRs. For instance, the mismatch between the IRR forecast and actual output and the high volatility of IRR power generation create a challenge for maintaining system frequency. ERCOT operators are forced to use more expensive ancillary reserves more frequently in order to maintain frequency at the desired level. To address this issue, ERCOT is improving IRR generation forecast tool for real-time operation while making changes to ancillary service requirement determination methodology to take into consideration the effects of IRRs.

In addition, IRRs in ERCOT are concentrated in the West/Far West Texas area; however, the major load centers are located in Dallas/Fort Worth and Houston areas. The amount of power transferred from remote locations to the load centers is limited by the existing transmission network capability. To fully utilize the available transfer capacity and take advantage of the latest IRR forecasting, ERCOT has implemented a real-time tool to monitor and enforce the latest transmission limits dynamically every 5 minutes. ERCOT also has implemented a set of special protection schemes (SPSs) and mitigation plans to deal with local congestion issues that constrain the IRR generation.

As the installed IRR generation capacity grows rapidly, ERCOT transmission system also needs to be expanded quickly to provide necessary transmission accesses for the new resources. In 2008, the Public Utility Commission of Texas (PUCT) ordered ERCOT to conduct several competitive renewable energy zone (CREZ) studies. Following the studies, the PUCT identified five CREZs in ERCOT based on the concentration of WGR potentials and financial commitments demonstrated by the developers. Later on, the PUCT further approved the construction of

2376 circuit miles of new 345-kV transmission to be completed by the end of 2013 to support higher wind penetration levels.

2   ERCOT Nodal Market

ERCOT nodal market includes congestion revenue rights (CRR) auction, a centralized day-ahead market (DAM), reliability unit commitment (RUC), and real-time security-constrained economic dispatch (SCED)—more details can be found in Chap. 2. DAM is a financial electricity market cleared in the day-ahead timeframe. The DAM clearing process determines unit commitments and awarded energy, ancillary service reserves, and certain types of CRRs which maximizes system-wide economic benefits given bids and offers submitted by the QSEs. The DAM scheduling also complies with network security constraint limits in addition to the usual unit commitment constraints.

RUC is a daily or hourly process to commit additional generation capacity on top of the self-committed capacity projected by the current operating plans (COPs) submitted by the QSEs to meet the forecasted demand. The DAM clearing is based on the voluntary energy offers and bids instead of the load forecast. The resource committed in DAM may not be sufficient to meet the real-time demand. Hence, the RUC process is needed to procure enough resource capacity to meet load forecast in addition to ancillary service capacity requirement.

There are three RUC processes used in the ERCOT nodal market:

- Day-ahead RUC (DRUC): DRUC runs once a day in day ahead for the next operating day. It is used to determine if additional commitments are needed to be made for the next operating day.
- Hourly RUC (HRUC): HRUC process is executed every hour. It is used to fine-tune the commitment decision made by DRUC based on the latest system condition.
- Weekly RUC (WRUC): WRUC process is an offline planning tool to commit resources with lead time more than a day. Its study period is configurable and could be up to 1 week.

During real-time operations, security-constrained economic dispatch (SCED) dispatches (normally every 5 minutes) online generation resources to match the total system demand while observing resource ramping and transmission constraints. The SCED process produces the base point for each generating resource. ERCOT uses these base points to deploy regulation up and regulation down, ancillary service to control system frequency and deploy responsive reserve, and non-spinning reserve ancillary services to solve potential reliability issues.

3   Short-Term Wind/PVGR Generation Forecasting and Current Operating Plan (COP)

Each hour, ERCOT produces and updates IRR generation forecasts for a rolling 168-hour in both system level and individual IRR level. The detail forecast types are described as follows:

- Short-Term Wind/PhotoVoltaic Power Forecast (STWPF/STPPF): It represents an hourly 50% probability of exceedance forecast of the generation in MWh per hour from each wind or solar resource.
- Wind-powered/PhotoVoltaic Generation Resource Production Potential (WGRPP/PVGRPP): The generation in MWh per hour from a wind or solar generation unit that could be generated from that resource allocated from the 80% probability of exceedance of the power potential.

The WGRPP is always lower than the STWPF for the same WGR since WGRPP is produced as 80% probability of exceedance of the forecast while STWPF is produced as 50% probability of exceedance. Similar relationship is also applied to STPPF and PVGRPP for the same PVGR.

At ERCOT, the wind and PVGR generation forecast is not directly used in the scheduling and RUC process. Current operating plan (COP) is a plan provided by a QSE reflecting anticipated operating conditions for each resource that it represents for each hour in the next seven operating days. The COP represents resource status, resource operational limits, and ancillary service schedules. The COP resource operational limits include high sustained limit (HSL) and low sustained limit (LSL). IRRs must keep their COP up to date to reflect the latest forecast and changes in capacity or resource status. For each of the 168 hours of the COP, a QSE representing a Wind-powered Generation Resource (WGR) or PhotoVoltaic Generation Resource (PVGR) must enter an HSL value that is less than or equal to the most recent forecast provided by ERCOT for that resource to show its production potential.

4   IRR Scheduling in DAM, RUC, and SCED

An IRR can participate in DAM, RUC, and real-time SCED, as detailed below.

4.1   IRR Scheduling in DAM

The QSE representing an IRR may participate in DAM for volunteer scheduling by submitting one of the following offers or bids in DAM:

- Submit three-part supply offer to sell energy for the physical IRR.
- Submit virtual energy-only offers to sell or bids to buy at the IRR settlement point.
- Submit CRR point-to-point obligation bids with source or sink at the IRR settlement point.
- Submit CRR point-to-point option offer with source or sink at the IRR settlement point.

The first option by submitting three-part supply offer is only available for QSE representing a physical IRR, and the other three options are available for all the QSEs.

Physical resources can participate in DAM for energy and ancillary service co-optimization by submitting three-part supply offer and ancillary service offer if they are qualified to provide the offered ancillary service. Currently, the IRRs are not qualified to provide ancillary service other than regulation down due to its variability and limited dispatchability. Thus, an IRR cannot submit ancillary service offer in DAM, and they can only submit three-part supply offer to DAM for energy clearing.

The wind/solar generation forecast is not directly used by DAM and RUC as input for their scheduling. Instead, DAM and RUC use the COP for the IRR in their scheduling. Since DAM is a financial market, only the HSL and LSL from the COP are used in DAM clearing without considering the resource status.

Instead, the resource status in DAM is determined by the existence of a valid three-part supply offer that can be used in clearing market. The HSL and LSL from COP are used in DAM co-optimization to determine energy and ancillary service award. The minimum of HSL and three-part supply offer largest offer quantity is used as the resource maximum dispatch limit in DAM, i.e., QSE can choose to partially participate in DAM energy clearing by offering partial capacity.

Figure 3.2 shows the average DAM hourly wind generation scheduling in August 2011 with the four options above. The DAM Wind in Fig. 3.2 represents the net energy injection at all the WGR settlement points from the awards of all the four options, i.e., the total sold quantity-total bought quantity. It can be observed that PTP Obligation award has the highest percentage in DAM wind scheduling. On the other side, the physical three-part supply offer awards in DAM are very low, representing less than 10% of the real-time wind output. Figure 3.2 also shows that in this scenario, wind generation in DAM is under-scheduled when compared to real-time output probably due to the uncertainty of wind generation.

4.2 IRR Generation Scheduling in RUC

Different from DAM, RUC is a reliability process, and all the physical resources including IRRs are required to participate in RUC by submitting a valid COP. Hence, both resource status and resource limits including HSL, LSL, and ancillary service schedules in the COP are used in RUC scheduling. The full capacity of HSL

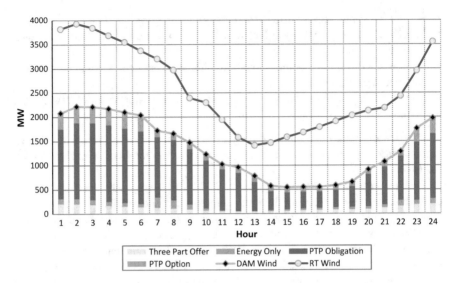

Fig. 3.2 Average DAM hourly wind generation scheduling in August 2011

is considered in RUC as maximum dispatch limit. Since IRRs are required to update their COP HSL constantly based on the most current forecast, RUC is able to take into account the most recent IRR output potential in its scheduling from the COP.

To verify the WGR COP HSL update accuracy, Fig. 3.3 is drawn to show the system total STWPF, WGRPP, and WGR COP total HSL for operating day of May 8, 2011, at the snapshot of DRUC execution time (14:30 on May 7, 2011). The real-time wind generation output is also included for comparison.

Figure 3.3 shows that the DRUC WGR COP HSL is between WGRPP and STWPF as expected which indicates that QSE followed the requirement to update the WGR COP according to the day-ahead STWPF. All the four lines are following the same real-time wind generation trend.

In RUC make-whole settlement, all QSEs that were capacity-short in each RUC will be charged for that shortage as RUC capacity-short charge. To determine whether a QSE is capacity-short, the 80th percentile IRR forecast for the IRR unit produced for the corresponding RUC is considered the available capacity of the IRR when determining responsibility for the corresponding RUC charges, regardless of its real-time output. In other words, even the high-value 50th percentile IRR forecast is used as HSL for RUC input to reduce RUC over commitment; the low-value, 80th percentile IRR forecast is used in RUC settlement to account for the IRR capacity.

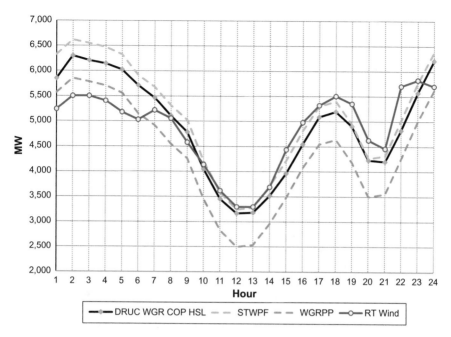

Fig. 3.3 Wind scheduling in DRUC

4.3 IRR Generation Scheduling in Real-Time SCED

In real time, all of IRRs are required to offer in their full capacity, i.e., their telemeter HSL equal to current net output of the facility. SCED dispatches generation every 5 minutes based on current IRR output to take care of net load variability. Once completed, SCED will issue each IRR a specific curtailment instruction for managing congestion if the IRR is required to produce the power output below its full capacity. An IRR also needs to limit its ramp rate to 10 percent per minute of its nameplate rating as registered with ERCOT when responding to or released from an ERCOT deployment.

Proxy energy offer curve may be used for IRR in SCED. Each generation resource provides either an energy offer curve (EOC) or an output schedule (OS) for ERCOT's use in determining how to dispatch the generation in the grid. An EOC or OS must be provided to ERCOT prior to the operating period and have restrictions on how they may be updated and changed. Due to no fuel cost for IRRs and government subsidies, most of IRRs submits zero or negative offers in the ERCOT nodal market. Consequently, IRRs are mostly price takers in the ERCOT market. However, under rare conditions, IRRs become marginal and set the market prices. If no EOC is provided, SCED creates a proxy energy offer curve for IRR with a -\$250/MWh when it runs to give the highest priority to IRRs while dispatching resource to meet the demand. Figure 3.4 shows an example of proxy offer created for a WGR without EOC. With EOCs specified for all resources, dispatch priority in

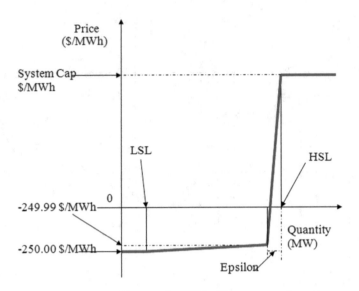

Fig. 3.4 Real-time proxy energy offer curve for WGR without energy offer curve

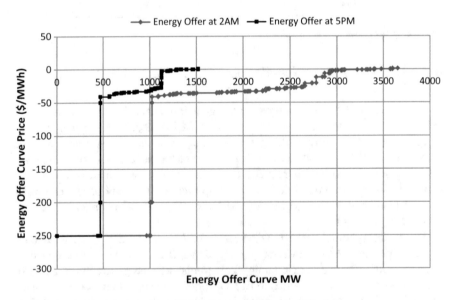

Fig. 3.5 Typical real-time system-wide wind aggregated energy offer curve

SCED is purely driven by supply prices for each generation resource when congestion does not exist.

Figure 3.5 compares the aggregated energy offer at 2:00 a.m. and 5:00 p.m. for a same day. The extended energy offer line at 2:00 a.m. was largely attributed to the higher wind power output in the night.

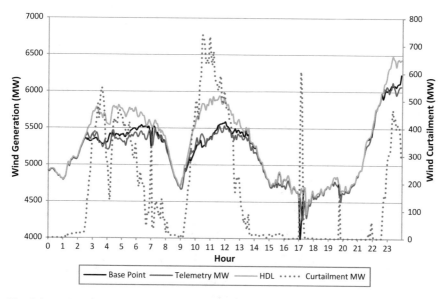

Fig. 3.6 Real-time system-wide wind generation curtailment

In order to better manage the IRR's real-time operation, Nodal Protocols have been revised to require ERCOT to send IRRs a curtailment flag indicating the base point dispatch instructions determined by SCED are below the high dispatch limit (HDL) used by SCED. The curtailment flag will be used by the settlement to determine when the IRR's base point deviation charge will apply. QSEs are also required to telemeter the real-time HSL of an IRR to ERCOT based on the following rules:

- If no curtailment, the telemetered HSL should equal to the current net output capability of the facility, i.e., IRR telemetry MW.
- If with curtailment, the telemetered HSL should equal to the estimated, non-curtailed output potential.

In this way, ERCOT is able to estimate the IRR's capacity more accurately during the real-time operation. In addition, ERCOT can apply a better base point deviation charge rules to IRRs, which will be discussed next.

To study the effectiveness of IRR HSL telemetry and curtailment flag, an operating day of August 30, 2010, is selected for the illustration. Figure 3.6 shows the real-time system-wide wind generation and curtailment for operating day of August 30, 2010. It can be seen that the WGR telemetry followed the base point very closely. The largest wind curtailment happened between 10 and 11 and is larger than 700 MW.

Figure 3.7 shows an individual WGR behavior following its base point and curtailment instruction on August 30, 2010. It shows that the WGR followed the SCED base point very closely all the time including curtailment period (when the

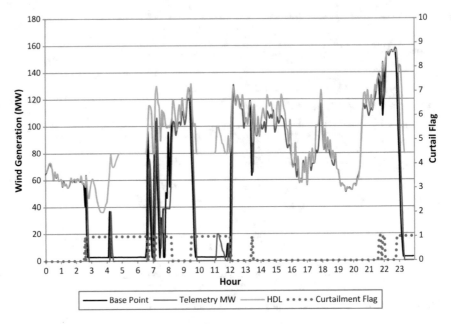

Fig. 3.7 Real-time curtailment of single WGR

curtailment flag equals to 1). It also can be observed during the curtailment period, the WGR correctly telemetered the non-curtailed output potential for its HSL as required by the Nodal Protocols instead of current output.

4.4   Base Point Deviation Charge for IRR

ERCOT charges an IRR a base point deviation charge if the IRR metered generation is more than 10% above its adjusted aggregated base point (AABP) when it is curtailed for congestion management. With regard to base point deviation charge, the key differences between IRRs and conventional resources are:

- Wider tolerance for IRR over-generation above base point (due to variability of energy sources)
- Only applied when the IRR is curtailed for congestion management
- No deviation penalty for under-generation below base point

The IRR must always take the necessary control actions, in its capability, to comply with base point dispatch instructions. If the curtailment flag signifying that the IRR has received a base point below the HDL used by SCED is set in all SCED intervals within the 15-minute settlement interval, then the base point deviation charge is calculated as

$$\text{BPDAMT } q, r, p = \text{Max } (\text{PR1}, \text{RTSPP } p) * \text{Max}$$
$$(0, \text{TWTG } q, r, p - \tfrac{1}{4}{}^* \text{AABP } q, r, p^* (1 + \text{KIRR}))$$

Otherwise

$$\text{BPDAMT } q, r, p = 0$$

where BPDAMT is the charge to the QSE for the IRR r at resource node p, for its deviation from base point for the 15-minute settlement interval; TWTG is the time-weighted telemetered generation of IRR at resource node p; KIRR is the percentage tolerance for over-generation of an IRR, 10%; RTSPP is the real-time settlement point price at resource node p; and PR1 is the minimum price to use for the charge calculation, $20.

4.5 Effect on Management of Congestion

In zonal market, management of congestion was done by issuing portfolio dispatch instructions for 15-minute interval based on forecasted information taken more than 20 minutes before the beginning of the 15-minute interval. Local congestion management was not very effective with portfolio balancing energy dispatches and zonal shift factors. Hence, ERCOT had to do resource-specific out-of-merit energy (OOME) dispatch instructions. Due to the inaccuracy of data, the possibility of issuing dispatch instructions only every 15 minutes, and the real-time volatility in wind, the major interface lines between the high wind area and load area had to be constrained to 60–70% of the allowable transfer limit to manage the flows below the transfer limit. This considerably affected the effective utilization of available wind energy.

In the nodal balancing market (SCED), resources are re-dispatched to manage congestion by issuing resource-specific instruction based on resource-specific shift factors, real-time generation, and real-time line flows. In addition, ERCOT has the flexibility to re-run SCED any time; thus, any significant change in IRR output could be handled by re-dispatching generation immediately. Hence, ERCOT is able to manage the flows in the major interface lines between the high wind area and load area below the limit by constraining to 80–90% of the allowable transfer limit. This has resulted in the better utilization of the available wind energy in the nodal market. The generation from wind has increased on an average by about 22% in the first 11 months (December 2010–October 2011) in nodal when compared to the last 11 months (December 2009–October 2010) in zonal mainly due to the better management of congestion.

4.6   Effect on Market Prices

ISOs need to procure more ancillary services to manage the uncertainties related to IRRs while maintaining reliable operation. More non-spinning reserve service is procured to account for wind and solar forecast error, and more regulation reserve service is procured to account for real-time variability in IRRs. These added costs to the loads for increased ancillary services are offset by the reduction in energy cost due to the low-cost energy from IRRs.

In ERCOT, WGRs have set the price system-wide during a few low load periods. Due to the high concentration of WGR in the west zone in ERCOT, WGRs set the prices in the west zone a considerable amount of time causing the average real-time price in the zone to be lower than the other zones. However, due to the variability of wind output, the average differences between day-ahead and real-time prices are much higher in west zones than in other zones. Increased penetration of IRRs reduces the expected hours of operation of non-IRR resources driving up the prices during peak load times for the non-IRR resources to recoup their capital cost.

5   Management of Generic Transmission Constraints (GTCs)

Generic transmission constraint (GTCs) is defined as a transmission constraint made up of one or more grouped transmission elements that is used to constrain flow between geographic areas of ERCOT for the purpose of managing stability, voltage, and other constraints that cannot otherwise be modeled directly in ERCOT's power flow and contingency analysis applications. Each GTC is designed so that the appropriate resources are controlled (as a result of the GTC) to protect against the stability phenomenon. Some GTCs have a closed interface (where all resources behind the interface have a similar effect on the stability phenomenon), while others are an open interface (where different resources have a different impact on the stability phenomenon). From GTC limit management perspective, a margin is desirable to be maintained so as to accommodate unaccounted factors when determining GTC limit, which include, are but not limited, to the deviations from base points. For those GTCs designated as IROLs (interconnection reliability operating limits), a tighter control is needed to avoid IROL exceedances in order to comply with NERC requirement.

5.1   Management of GTC Limits

Today, SCED dispatches units in the most optimal manner in order not to exceed the GTC limit by considering offer and shift factor from resources—see Fig. 3.8. If all

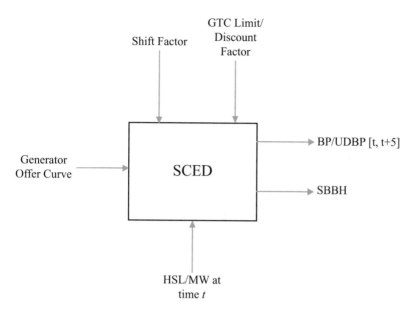

Fig. 3.8 Management of GTC

units have the same SF (shift factor) on GTC elements (closed interface), then units with the highest offers are curtailed down to where interface flow is within limit. It does this by setting SBBH (SCED base point below HDL) flag for the marginal units and requiring them to not exceed (or follow) their UDBP (updated desired base point). However, other infra-marginal units are not required to follow their base points since their SBBH flag is not set. Those IRR units without SBBH flag to be set can increase output above UDBP, sometimes very quickly, if wind speed or irradiance allows them to do so.

Figure 3.9 shows such an instance when the actual Panhandle GTC power flow is exceeding its limit. SCED has been running to produce a series of base points or UDBP every 5 minutes as the orange line. However, those non-marginal IRR resources, which were not instructed by ERCOT to curtail their power output, continued to ramp up their power shown as the gray line. As a result, the actual interface power flow (red line) was greater than the limit (blue line).

To operate the grid in a conservative way, the current approach is to bind and dispatch GTC flow to a certain lower limit (as a percentage of actual limit) due to potential for IRR units that do not have SBBH flag to ramp up. This percentage is lower for GTCs that are IROLs due to increased reliability implications of exceeding the limit.

However, there are some difficulties in selecting the appropriate percentage. First, controlling to a higher percentage of the limit will not be sufficient to avoid exceedances in the case when IRRs ramp up (reliability risk). Second, controlling to a lower percentage of the limit may lead to excessive curtailment of IRRs (efficiency loss) and more binding intervals for all units behind the GTC. If assessed

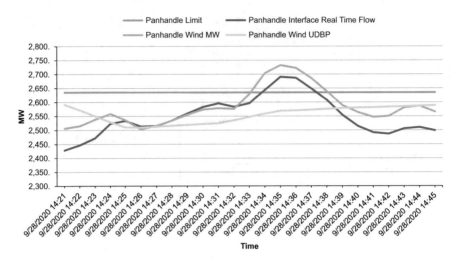

Fig. 3.9 Illustration of ramp up for uncurtailed IRRs causes GTC limit to be exceeded

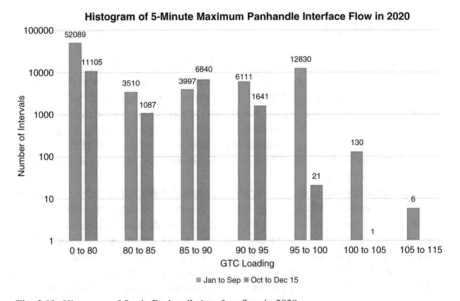

Fig. 3.10 Histogram of 5-min Panhandle interface flow in 2020

as IROL, a lower percentage may be needed per operation procedures to avoid NERC reporting.

Figure 3.10 shows the histogram of 5-min Panhandle interface flow. From January to September 2020, ERCOT controlled to 90–95% of the actual limit for the vast majority of intervals, i.e., there were a total of 136 instances of exceedance of Panhandle GTC limit between January and September 2020. Now that the Panhandle GTC is an IROL, ERCOT has controlled to 85–90% of the limit, in order to prevent any exceedances which in turn result in more binding intervals and curtailment of wind resources at the Panhandle.

5.2 Non-to-Exceedance (NTE) Method

To achieve a balance between efficiency and conservativeness, a new method, referred to as non-to-exceedance (NTE) method, was innovated. Essentially, when NTE is activated, IRR units behind a binding GTC with a considerable impact over the GTC flow should not exceed their SCED base point.

The key parameters for the implementation of the NTE method include:

– *Triggering condition:* Trigger conditions being considered are:

1. When GTC loading is above a threshold
2. When GTC constraint is activated/binding in SCED

– *Discount factor:* Reliability limit will be discounted in a less conservative manner, closer to 1.0.
– *Unit impacts:* IRR units with a shift factor greater than 2% on a GTC cannot exceed their SCED base point.

NTE can bring a number of benefits to the grid operations. First, a NTE method should allow ERCOT to bind at a higher limit, resulting in more MW production from IRRs behind IROL when binding and allowing for fewer time intervals of low prices behind the constraints. Second, it keeps ERCOT from exceeding SOLs and IROLs. Third, it provides a more efficient way of meeting reliability requirements, especially when GTC limit constraints are binding. Fourth, it offers additional advantages to real-time operations, which include (1) less operator attention and (2) a greater certainty in managing GTC power flow.

Figure 3.11 demonstrates another example of potential benefits brought by a NTE method. The dot purple line represents the real-time GTC limit before applying a reliability discount factor. The blue solid line was the actual GTC interface power flow under the non-NTE control paradigm where a violation of the limit can be observed. To operate the grid more conservatively, a lower reliability discount factor needs to be applied, shown as the blue dot line, and the periodic adjustment to this discount factor is also required. However, the resulting interface power flow (black dot line) was still overconstrained between 14:35 and 14:40. Under the new NTE paradigm, a constant reliability discount factor, close to 1.0, is needed, shown as the yellow dot line. The corresponding interface power flow (green solid line) followed this instruction very closely, and thus, it was guaranteed that the controlled power flow (green solid line) was below the target (yellow dot line).

Figure 3.12 shows another example of how the proposed method can improve the management of the interface power flow under the limit. The blue line denotes the interface flow limit. The power flow without a NTE is shown in the green line, and

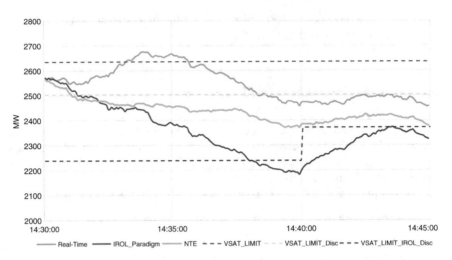

Fig. 3.11 Illustration of NTE

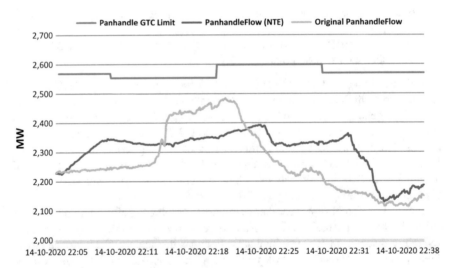

Fig. 3.12 Another example of control of NTE method

the resulting power flow is represented as the brown line after the NTE is applied. It clearly demonstrates that starting at 22:11, the original interface flow was approaching the limit quickly, which could cause a reliability concern as the safety margin is reduced. Afterward, the original power flow decreased at a fast rate because operators managed the power flow in a more conservative way by adopting a lower discount factor. This caused more wind power to be spilled off unnecessarily. In contrast, as the wind speed stayed high from 22:05 to 22:31, NTE method attempted to maintain a relatively constant margin between the imposed limit and the interface flow.

NTE could also have an impact over the locational marginal price (LMP) as it allows more power to flow out of the congested area. As a result, the marginal unit could change. Table 3.1 shows the comparison of LMP between those methods without and with NTE method implemented. NTE would always result in an increase in LMP due to a less conservative operation, and the change could be significant if the difference between the resulting power flow was also large. It clearly shows that a NTE method can benefit both the grid reliability and the price formulation.

5.3   Implementation of NTE Concept

While the NTE method is efficient in the better control of IRRs behind a GTC, other two alternatives exist—as shown in Table 3.2. The first option (option 1) is to require

Table 3.1 Impact over LMP

LMP ($/MWh)	14OCT2020: 22:05:18	14OCT2020: 22:10:19	14OCT2020: 22:15:17	14OCT2020: 22:20:18	14OCT2020: 22:25:16	14OCT2020: 22:30:18
Averaged LMP- NTE	-17.2	-18.4	-18.5	-20.6	-29.3	-29.3
Averaged LMP-original	-18.7	-18.7	-18.7	-31.5	-31.2	-31.2
Delta	1.5	0.3	0.3	10.9	1.8	1.9

Table 3.2 Alternatives to NTE

	Option 1	Option 2	Option 3
Description	IRR cannot exceed UDBP regardless of SBBH	To set SBBH flag for all IRR resources behind constraint when the GTC limit is binding (short-term solution)	To use 5-min forecast to calculate NTE for IRR resources when the GTC limit is binding (long-term solution)
Additional telemetry data	No	No	Yes
System changes needed	No system changes needed	Without significant ERCOT system changes and using existing interfaces so as to avoid QSE changes	More ERCOT system changes needed
Protocol changes needed	Extensive	Minimum	In-between
Efficiency	Lower	In-between	Higher

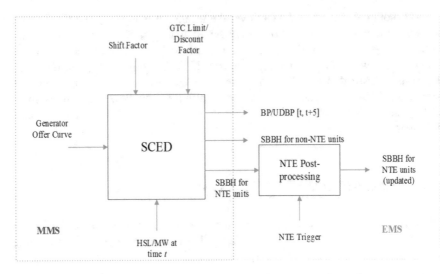

Fig. 3.13 Implementation of NTE method (option 2) (*MMS* market management systems, *EMS* energy management system)

all of IRRs not to exceed their UDBP. While it is simple in the design and does not require any system changes, it also suffers from a lower efficiency such that IRRs have to spin off power output over all of the time regardless of whether there is a reliability risk to the GTC limit. The third option (option 3) is more complex in the design as it utilizes the forecast for the next 5 minutes when determining UDBP. This especially benefits those infra-marginal units since those units can produce the power at their full potential when being an economical resource. It has a gain over the efficiency, but it requires much more complicated system changes. The option to be implemented (option 2) is a compromise between other two options, with a balance between the efficiency and implementation complexity.

A high-level description of the implementation of the NTE method is shown in Fig. 3.13. In this regard, no changes are needed to market management system (MMS) as the SCED still takes the existing input and produces the base points and the curtailment flag. A new post-processing step will be added once the NTE method is implemented. This step will validate when the NTE is activated and then override the curtailment flag for all of IRRs behind the corresponding GTC with a shift factor greater than 2%. In this case, the curtailment flag will be set either when a curtailment is needed or when a NTE method is enabled, thus without requiring additional telemetry data.

To optimize the performance of the NTE method, cautions need to be excised regarding when to activate a NTE method. First, for open-loop GTCs, shift factors of IRRs differ. Operators will only activate a constraint for consideration in SCED if there is at least one unit with an absolute shift factor greater than 0.02. Figure 3.14 shows the distribution of shift factors for one of the GTCs at the ERCOT grid. IRRs behind this GTC have shift factor ranging from 0.4 to 1.0. This indicates that all of

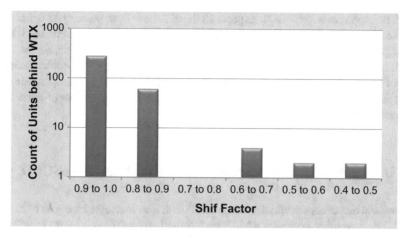

Fig. 3.14 Distribution of shift factors

Table 3.3 Comparison of curtailment at different maximum GTC loading

	Curtailment for maximum GTC loading at 85% (MWh)	Curtailment for maximum GTC loading at 90% (MWh)	Curtailment for maximum GTC loading at 95% (MWh)
W/o NTE	1,148,746	870,644	642,247
W/ NTE (activated when GTC loading >85%)	1,217,132	939,030	710,633

IRRs have a significant impact over the GTC interface flow. Otherwise, if an IRR has a smaller shift factor, it is not effective in controlling this resource in order to constrain the GTC power flow.

In addition to avoidance of exceedance, the NTE method can result in a more efficient management of GTC flow, i.e., less curtailment, if activated when GTC loading is greater than 85%. For example, NTE could have reduced curtailment on the Panhandle GTC from 870,644 MWh (or 7.29% of total energy) (a tighter control implemented today) to 710,633 MWh (or 5.87% of total energy) between January and September 2020, presuming discount factor can be increased to 95%. Also, reduction in the curtailment would be greater if controlling to 80% due to IROL today. Pricing is also impacted in fewer hours if EROCT is able to bind the constraint in fewer intervals. The simulation results show that the extra spin-off due to NTE (activated when GTC loading is greater than 85%) is 0.56% of total wind energy or 1.21% of wind energy during the period when NTE is activated—see Table 3.3.

Current SBBH is issued to resources in cases where the LMP is aligned with the base points on the resource energy offer curve, and the expansion of SBBH to the NTE approach will also include issuing an SBBH to resources in cases where the LMP is not aligned with the base points on the resource energy offer curve. Once

NTE is implemented, ERCOT expects IRRs to follow base point. The benefits of NTE can be maximized only when each IRR behind the GTC follows its base point. If there is a large over-generation above the base point for an IRR behind the GTC when SBBH flag is set, this may threaten the grid reliability and thus could run into a risk of being instructed to be disconnected from the grid. In addition, operators may have to reduce the discount factor to accommodate the expected over-generation above base points.

6   Incorporation of 5-Minute Wind/Solar Ramp into SCED

Renewable resources are variable and intermittent, thus increasing the grid balancing need. The prevailing approach when scheduling the renewable resources in SCED is to assume that wind or solar generation for the next 5 minutes will remain the same as the power production observed at the time of SCED execution. However, if wind or solar generation changes, it could exhaust the regulation up resources or cause the frequency control to be deteriorated. One example is presented in Fig. 3.15. As wind and solar continued to drop for three consecutive SCED intervals, all of regulation up resources were deployed, but the frequency was still not recovered to 60 Hz. Eventually, operators had to add a SCED manual offset to request more generation capacity to be dispatched. While this type of manual action can relieve the problem to some extent, it creates a burden to operators, thus rendering itself as an inefficient approach.

In the absence of a way of accounting for the ramping of wind or solar generation in real time, SCED dispatches uncurtailed solar units to their HSL. This means that SCED assumes output of uncurtailed solar units will persist for the next 5 minutes at the unit's HSL that was telemetered at the start of the SCED interval.

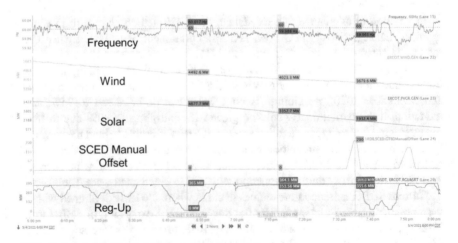

Fig. 3.15 Impact of solar and wind ramp

The difficulty can be better explained by another example. Assume a scenario at 5:00 p.m. when non-solar generation produces a power at 45,000 MW and the telemetered solar output and HSL is 3500 MW. At 5:01 p.m. when the SCED runs, GTBD is set to 48,500 MW if assuming solar generation will not ramp for the next 5 minutes, and no regulation has been deployed and an integral ACE is set at 0 MW. Assuming there is no curtailment, solar resources will be given base points equal to the HSL prior to the SCED run, which is 3500 MW. If from 5:00 to 5:05, actual solar power production ramps down by 300 MW, telemetered solar power generation will be 3200 MW, while non-solar generation produces 45,000 MW. In order to maintain a grid balance between the demand and generation, regulation deployment is used to make up for the gain or loss of solar power production, but depending upon the circumstance, the frequency recovery could potentially be delayed.

6.1 Generation to Be Dispatched (GTBD)

ERCOT has implemented a new approach to incorporate the 5-minute wind and solar generation ramp into the generation requested for SCED, GTBD (generation to be dispatched). GTBD is the short-term 5-minute prediction used in SCED. GTBD is the sum of the base points dispatched in SCED. Typically, it is calculated by summing up the current actual generation from all the generation resources which have been previously dispatched by SCED. The formula for GTBD is given as follows.

$$
\begin{aligned}
\text{Generation To Be Dispatched (GTBD)} = {}& \text{Total Generation} \\
& + K_1{}^*10^*\text{System Load Frequency Bias} \\
& + K_2{}^*[(\text{net non} - \text{conforming Load}) - (\text{net filtered non} - \text{conforming Load})] \\
& + K_3{}^*5^*\text{PLDRR} + K_4{}^*\text{Regulation Deployed} + K_5{}^*\text{ACE Integral} \\
& - K_6{}^*5^*\text{PWRR} + K_7{}^*5^*\text{DCTRR} - K_8{}^*5^*\text{PSRR}
\end{aligned}
$$

$$(3.1)$$

where PLDRR is the predicted load ramp for the next 5 minutes, PWRR is the predicted wind generation ramp for the next 5 minutes, DCTRR is the predicted DC tie ramp for the next 5 minutes, and PSRR is the predicted solar generation ramp for the next 5 minutes.

K_1, K_2, and K_3 are values that can be tuned to predict GTBD. This allows the ability to account for season bias possibilities, i.e., a higher K_3 will be set for seasons that tends to be more volatile. By default, $K_1 = K_2 = 0$. Therefore, the calculation encompasses the total generation which should essentially be the total load in ERCOT at the start of the SCED interval, and the K_3 term is the projected load change in 5 minutes. Since May 2013, the K_4 configurable factor was added together with the rolling average of regulation deployed in the past X minutes where X can be a tuned value. When "Regulation Deployed" is calculated, it is subject to a constraint

$$|\text{Regulation Deployed}| \leq \text{Max Regulation Deployed Feedback} \qquad (3.2)$$

Since the K_4 factor could change quickly between SCED intervals because regulation should reset at the start of a new SCED interval, another term was added to consider longer-term resources performance and accuracy regulation feedback. Since December 2016, the K_5 configurable factor was added together with the integral area control error (ACE) calculated in the energy management system (EMS). The ACE algorithm subtracts the actual frequency in Hz from the scheduled system frequency (usually 60 Hz) and multiplies this value by the frequency bias constant in MW/0.1 Hz. It provides an ability to integrate the raw ACE over Y seconds for use in the ACE Integral equation, where the variable Y can be configured. The ACE Integral is also limited by maximum allowed feedback.

$$|\text{ACE Integral}| \leq \text{Max Integral ACE Feedback} \qquad (3.3)$$

It is worthwhile noting that the regulation resources could be deployed as the wind or solar generation moves within an interval. By adding the K_4 term, GTBD accounts for the change in wind or solar generation within a SCED interval. However, it is the reaction to this change but not a look-forward dispatch to re-position the dispatchable fleet for the anticipated change in the next 5 minutes.

To solve this problem, the updated formula (3.1) to calculate GTBD value includes a predicted 5-minute solar ramp (PSRR) component. A 5-minute solar ramp rate can be calculated from the Intra-hour PhotoVoltaic Power Forecast (IHPPF) and the Short-Term PhotoVoltaic Power Forecast (STPPF), with more details provided in the following subsection. A manual toggle is available to select one of two methods as primary source for setting the PSRR value to be used.

6.2   Forecast of 5-Minute Solar Ramp

Three methods have been proposed to calculate 5-minute solar ramp, PSRR, used in GTBD, which are described as follows.

First Method
The current solar HSL and the forecast for the current interval are denoted as SG0 and FSG0, respectively. The solar forecast for the next 5 minutes is FSG1. As shown in Fig. 3.16, the prediction of solar ramp for the next 5 minutes is calculated as

$$\text{PSRR} = (\text{FSG1--FSG0})/5 \qquad (3.4)$$

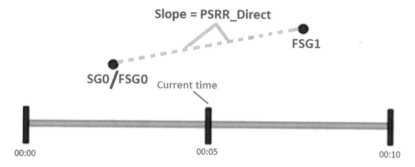

Fig. 3.16 Calculation of 5-minute solar ramp (first method)

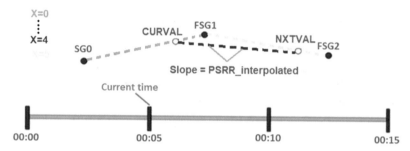

Fig. 3.17 Calculation of 5-minute solar ramp (second method)

Second Method
In the second method (see Fig. 3.17), to calculate 5-minute solar ramp, the prediction of solar power production is first interpolated, aligned with the starting time of the solar ramp to be predicted. The solar power production at the current time is the linear interpolation between the previous 5-minute power observation and the forecast for the current time interval

$$\text{CURVAL} = (\text{FSG1} - \text{SG0}) * t_X/5 + \text{SG0} \tag{3.5}$$

Similarly, the solar power production at the time of 5-minute past the current time is

$$\text{NXTVAL} = (\text{FSG2} - \text{FSG1}) * t_X/5 + \text{FSG1} \tag{3.6}$$

The 5-minute solar ramp is given by

$$\text{PSRR} = (\text{NXTVAL} - \text{CURVAL})/5 \tag{3.7}$$

where the configurable t_X is the time between t and $t + 5$, representing the delay of the receipt of solar forecast past the top of each 5 minutes, SG0 is the averaged HSL between t and $t + 5$, FSG1 is the forecast for $t + 5$, and FSG2 is the forecast for $t + 10$.

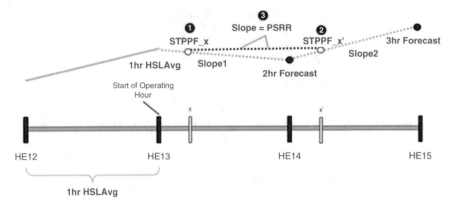

Fig. 3.18 Calculation of 5-minute solar ramp (third method)

Third Method

The third method utilizes the hourly solar forecast rather than the 5-minute counterpart as shown in Fig. 3.18. The rates of change in solar power generation for the current hour and the next hour are given by

$$\text{Slope1} = (\text{Forecast } (t + 1 \text{ Hour}) - 1\text{hr HSLAvg})/12 \tag{3.8}$$

$$\text{Slope2} = (\text{Forecast } (t + 2 \text{ Hour}) - \text{Forecast } (t + 1 \text{ Hour}))/12 \tag{3.9}$$

The solar power production at the current time (t_X shift from the top of the current hour) and 1 hour shift ($t_{X'}$) is

$$\text{STPPF\_}x = 1\text{hr HSLAvg} + (\text{Slope1} * t_X) \tag{3.10}$$

$$\text{STPPF\_}x' = \text{Forecast } (t + 2 \text{ Hour}) + (\text{Slope2} * t_X) \tag{3.11}$$

The predicted solar ramp is given by

$$\text{PSRR} = (\text{STLF\_}x' - \text{STLF\_}x) * 5/12 \tag{3.12}$$

The comparison of the forecasting performance for the first and third methods, evaluated in mean absolute error (MAE), is given in Table 3.4, when using the data from April 23, 2021, to May 3, 2021. Persistence method assumes 0 MW of solar ramp for the next 5 minutes. It can be seen that the solar ramp based on intra-hour solar forecast (first method) yielded the best performance when evaluated for all of periods and large solar ramp events, while the solar ramp prediction derived from short-term solar forecast (third method) was less accurate than persistence. This observation can be also confirmed by the histogram of PSRR error shown in

Table 3.4 Performance of 5-minute solar ramp prediction

Performance Metric	Persistence (4/23 – 5/3)	Intra-Hour Solar Forecast (4/23 – 5/3)	Short-term Solar Forecast (STPPF) (4/23 – 5/3)
MAE (MW)	72.10	60.99	79.32
MAE (MW) when 5-Min. Solar Ramp > 100 MW	185.77	128.30	179.32
MAE (MW) when 5-Min. Solar Ramp < -100 MW	189.95	122.95	197.34

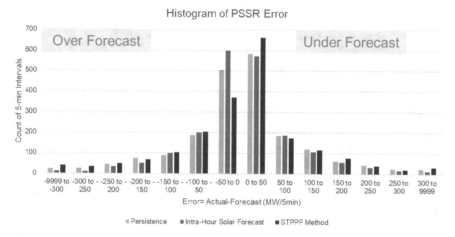

Fig. 3.19 Histogram of PSRR errors

Fig. 3.19. Intra-hour solar forecast can significantly reduce the PSRR errors at the tailed part of the histogram. This is very valuable since these large forecast errors present the largest reliability risk to the grid operations, which need to be mitigated most.

This PSRR times five and a configurable factor K_8 will be used in GTBD to capture the forecasted 5-minute solar ramp. This addition would help SCED to take into account the potential 5-minute solar ramp in its 5-minute base points and would take the burden off of regulation service to cover the 5-minute gain or loss of generation resulting from variations in solar irradiance and instead dispatch the energy more economically. This will also aid in reducing frequency recovery duration following events that occur during times with significant solar up and down ramps. Although there will be inherent forecast error in the solar forecast, this change will give a better indication to SCED via GTBD of the level of solar generation that will be available in the next 5 minutes, as compared to the current assumption that the solar generation will persist at the same level as at the beginning of the 5-minute interval. This 5-minute forecast in GTBD would help SCED better

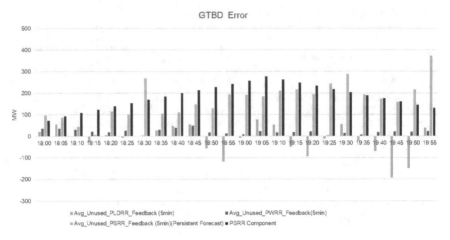

Fig. 3.20 Contribution to GTBD errors

dispatch base points to generation resources to account for the ramping of solar generation in the next 5 minutes.

The effect of incorporating solar ramp into GTBD can be demonstrated by one example given in Fig. 3.20. During the sunset hours, load forecast errors (PLRR) may occur occasionally, but less than 200 MW. At the same time, wind forecast was much more accurate, which only resulted in less than 50 MW of the wind forecast error (PWRR). The largest contribution to GTBD errors came from the solar ramp. As the solar generation ramped down, if the persistence method were used, the PSRR error would be between 100 MW and 400 MW. After the inclusion of PSRR in GTBD, the PSRR (blue bar) was better aligned with the actual solar ramp (green bar). Except for five intervals (18:00, 18:30, 19:30, 19:50, and 19:55), the solar generation ramp was over-forecasted. This clearly shows the benefit of including PSRR into GTBD, which only requires a small amount of GTBD error to be compensated by the deployment of regulation resources.

7   Conclusions

This chapter discusses the customized market solutions implemented at ERCOT for the new nodal market enables Wind Generation Resources (WGRs) or PhotoVoltaic Generation Resources (PVGR), which are collectively referred to as Intermittent Renewable Resource (IRR) here, to realize their asset's value and ensures the system reliability at the same time.

In a nodal market, an IRR can participate in DAM and RUC by using the latest hourly forecast to submit current operating plan (COP) to reflect the projected capacities and statuses for the next 168 hours.

In real-time operation, the intermittence and variability of IRRs could impose a challenge to manage GTCs causing the exceedance of the transmission power transfer limit. As the penetration of IRRs is increasing, it is very important to request all of IRRs to follow their non-to-exceedance base points rather than a subset of IRRs to be curtailed, especially when the GTC constraint is binding.

In real time, SCED also needs to know how much the generation capacity is dispatched. IRRs, especially solar generation resources, could introduce large variations for the next 5 minutes. To compensate for this variation, 5-minute wind and solar ramp was incorporated into the SCED generation request. ERCOT's experiences show that doing that can significantly improve the frequency control performance while reducing the need for regulation resources.

Chapter 4
Ancillary Services (AS) at ERCOT

1 Overview

Ancillary services in ERCOT currently include responsive reserve service, regulation service, non-spinning reserve service, black start, and emergency interruptible loads. The latter two services are used in emergency and system recovery conditions, while the former three services are used to balance net load[1] variability and support frequency after generation outages.

1.1 *Responsive Reserve Service (RRS)*

Responsive reserve service (RRS) bundles two distinct functions within one service. This reserve is used for frequency containment, i.e., to arrest frequency decline after generator trip, and as a replacement reserve to restore the depleted responsive reserves and bring the frequency back to 60 Hz. Until recently (June 2015), ERCOT procured 2800 MW of RRS for every hour in a year.

Recently, due to the changing generation mix in ERCOT, the methodology for determining RRS has changed from procuring a constant amount of 2800 MW for all hours to determining necessary amounts of RRS dynamically based on expected system inertia conditions. Of this amount, 60% could be provided by interruptible load resources with automatic under-frequency relays. The relays are activated within 0.5 seconds, if system frequency drops to 59.7 Hz or lower. These interruptible loads providing RRS are usually large industrial loads. The remainder of RRS is provided by generation resources and is deployed autonomously through governor response (as containment reserve).

---

[1] Net load is defined as system load minus wind and solar power production.

© The Author(s), under exclusive license to Springer Nature Switzerland AG 2023
P. Du, *Renewable Energy Integration for Bulk Power Systems*, Power Electronics
and Power Systems, https://doi.org/10.1007/978-3-031-28639-1_4

1.2   Regulation Reserve Service

Regulation reserve service is a restoration service, used to restore frequency back to 60 Hz after a disturbance as well as to balance out intra 5-minute variability in net load [1]. Resources providing regulation service respond to an automatic generation control (AGC) signal from ERCOT every 4 seconds. The regulation reserve requirements are determined separately for regulation up and regulation down for each month, hours 1 through 24 (i.e., 24 values per month). The requirements are based on regulation service deployments for the same month in the previous 2 years and a certain percentile for the 5-minute net load variability for the same period. ERCOT also can increase regulation requirements based on historic exhaustion of regulation reserves as well as make adjustments to account for newly installed wind and solar generation capacity that is not yet captured in the historic evaluation period.

Somewhat counterintuitively, regulation requirements in ERCOT are trending down with increasing installed wind and solar generation capacity. This can be due to the increase of geographical dispersion of wind and solar generation resources as well as the continuous fine-tuning of AS methodology for determining the requirements.

Fast-responding regulation service (FRRS) was introduced in 2013 as a subset of regulation service to allow resources to provide fast frequency support to the system. Resources providing FRRS are required to respond to a separate AGC signal as well as to a local frequency trigger (currently set at 59.91 Hz). Each resource with FRRS obligation should provide its full response within 1 second.

1.3   Non-spinning Reserve Service

Non-spinning reserve service (Non-Spin) is a service to provide support within 30 minutes through online and/or offline resources. Non-Spin may be deployed to react to the loss of a generator, to compensate for net load forecast error, and to address the risk of large net load ramps or when low amount of generation capacity is available in SCED. Historically, the need for Non-Spin has occurred during hot or cold weather, during unexpected weather changes, or following large unit trips to replenish deployed reserves.

2   Regulation Services

Regulation service is an ancillary service that provides capacity that can respond to signals from ERCOT within seconds to respond to changes in system frequency. There are two different types of regulation services: regulation up and regulation down, commonly referred to as reg-up and reg-down.

Wind generation adds value to the electric power industry by diversifying generation portfolios, reducing fuel costs, hedging against fluctuating gas prices, and providing a sustainable source of clean energy [2–6]. However, the variability and uncertainty of wind generation raise serious operational challenges, such as deteriorated forecast accuracy and increased needs for ancillary services to ensure grid reliability and security [2–6]. Regulation services (RSs) are part of the ASs necessary to maintain the reliable operation of the transmission system. A comprehensive discussion of the AS market can be found in [7, 8]. As variable resources like wind generation continue to grow, the definitions, rules, and methods concerning AS also evolve [9, 10]. Many recent studies have focused on how to provide spinning or "load following" reserves [11–20]. In efforts to quantify the RS requirements, investigations have been confined to using overly simplified models to estimate the RS requirements [21, 22]. However, in actual grid operation, detailed models and historical operational data are needed to quantify the RS requirements and to verify their effectiveness.

Two new algorithms were developed at the Electric Reliability Council of Texas (ERCOT) for determining the RS requirements to cope with the uncertainties and variations brought by high penetration of wind generation resources. ERCOT is an independent system operator (ISO) serving over 23 million customers in Texas. ERCOT leads the United States with over 32 GW of installed wind capacity by the end of 2021. In 2021, the highest instantaneous penetration level of the wind generation was 66.47%. However, as a single balancing authority (BA) without synchronous connections to its neighboring systems, ERCOT relies purely on its internal resources to balance generation shortages and variations. Therefore, it is critical that the derived AS requirements meet the reliability needs.

The first algorithm is a novel short-term wind forecast algorithm that can be integrated into the look-ahead dispatch process to improve the calculation of locational marginal prices (LMP) and generation schedules. The second enhancement is the development of a multi-time-scale method for quantifying the regulation service requirements considering the combined impact of wind variations on governor response, AGC, and economic dispatch results.

In areas with wind penetration above 20%, look-ahead dispatch algorithms [23–25] are commonly used to decide LMP and power generation schedule to cope with the variability and uncertainty of wind. At each time interval, the look-ahead dispatch executes economic dispatch for the next hour with a 5-minute interval. Compared with the traditional process, in which only 1 economic dispatch is executed for the next 5-minute interval [26], 11 more dispatch intervals can be obtained by the look-ahead dispatch process to account for the inter-temporal constraints in the scheduling process. Therefore, the possibility of violation of security constraints and price volatility is significantly reduced. In addition, manually dispatching out-of-merit-order generators can be avoided. To further improve the performance of the look-ahead dispatch process, a novel short-term wind forecasting algorithm is developed and integrated into the real-time security-constrained economic dispatch (SCED). The algorithm combines the persistent forecasting model with the weather numerical prediction model (WNPM) so that a better performance can be achieved. By updating the wind forecast every 5 minutes using the proposed algorithm, the look-ahead dispatch achieves better results on LMPs and more efficient generation schedules.

The second enhancement ensures a sufficient amount of RSs are purchased to cope with the variability in each dispatch interval. The method is proposed to determine the regulation requirements by explicitly considering the net load in the context of economic dispatch. A detailed simulation model of the ERCOT system is used to assess the adequacy of the procured regulation services. Note that the model used in this study is able to simulate realistic operation scenarios across different time scales, i.e., including the governor response, AGC, as well as economic dispatch, and the performance is thoroughly examined by the standard control performance system score and by the proposed performance metric—exhaustion rate of regulation resources. In addition, a year-long historical field data was collected at ERCOT to validate the performance of the proposed method.

2.1   Short-Term Wind Generation Forecasting

At ERCOT, at the beginning of each hour, a wind forecast for a rolling 48-hour is used as the input to the real-time SCED [26].[2] Previously, a look-ahead dispatch mainly includes load forecast. For example, the Midcontinent Independent System Operator (MISO) used a time-coupled multiple-interval dispatch to explicitly ensure that the ramping capability is cost-effectively obtained [23]. While the implementation of a time-coupled multiple-interval dispatch can potentially reduce the production cost due to the better pre-positioning of generation resources [27], its primary enhancement is the improved reliability. Such a look-ahead real-time dispatch (RTD) has also been developed at ERCOT [28–30], which involves multiple time intervals and considers temporal constraints for enabling additional resources to contribute to the real-time price formation and for expanding access to the real-time energy market. The dispatch is executed every 5 minutes. Note that each 1-hour study period is divided into 12 5-minute time intervals.

An example at ERCOT is used to demonstrate the need of an accurate short-term wind forecast for the RTD process. Figure 4.1 shows the actual system net load,[3] the actual SCED system lambda, and the simulated RTD system lambda (e.g., the shadow price that bounds the power balance constraint) between 10:00 and 10:50 for an operating day in 2015. A price spike $1046/MWh occurs at 10:40 to meet the ramping constraint when using the single time-interval SCED. Using the short-term

---

[2]ERCOT is running an LMP-based market, which includes a centralized day-ahead market (DAM), reliability unit commitment (RUC), and real-time SCED. During real-time operations, SCED dispatches (normally every 5 minutes) online generation resources to match the total system demand while observing resource ramping limits and transmission constraints.

[3]The net load is defined as the "the aggregate customer load demand minus the aggregate wind generation output" and is the variable that the power system operators use to dispatch generation.

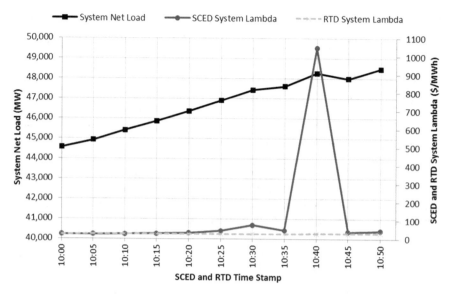

Fig. 4.1 Comparison of results from RTD and SCED

load and wind forecast, the simulated RTD system lambda is reduced to $28.7/MWh, showing that accounting for both the wind and load variations can effectively reduce the total production cost when large ramping events occur. The RTD system lambda is solved after two runs: a dispatch run and a pricing run. To accurately model the ramp rate, the dispatch run uses a mixed-integer quadratic programming (MIQP) algorithm to determine the optimal dispatch schedule. In the pricing run, all RTD study intervals are decoupled and solved as a single-interval SCED using a quadratic programming (QP) algorithm with resource dispatch limits determined from the MIQP solutions. Therefore, it can be more efficiently used to address reliability considerations, leading to a more economic dispatch solution.

A short-term wind power forecast (STWPF) can be produced either by the WNPM or using advanced statistical methods [31–33]. Including the STWPF in the look-ahead period of study accounts for the wind variations and reduces the error in the LMP calculation. Therefore, a short-term wind forecast model is developed at ERCOT. Wind farm outputs are forecasted for the first 5-minute interval, the same as the current telemetered wind power. For every future 5-minute intervals, the forecast is calculated as a function of the last three average telemetry outputs, the current telemetry output, and the forecast output obtained using the WNPM for the current hour. Mathematically, the process is represented by

$$
\begin{aligned}
\gamma_{5minute}(t+5) = {} & A \cdot \alpha + B\,\gamma_{5minute}\,(t-5) + C\,\gamma_{5minute}\,(t-10) \\
& + D\,\gamma_{5minute}\,(t-15) + E\,\gamma_{hour}
\end{aligned} \tag{4.1}
$$

where $\gamma_{5minute}\,(t+5)$ is the wind forecast for the future 5-minute interval; α is the current wind generation; $\gamma_{5minute}(t-5)$, $\gamma_{5minute}(t-10)$, and $\gamma_{5minute}(t-15)$ are the

Table 4.1 Performance comparison between the persistent method and the proposed method

Interval	MAE of persistent method (MW)	MAE of method used by ERCOT (MW)
5 minutes	58	47
10 minutes	101	81
15 minutes	142	114
20 minutes	182	149
25 minutes	221	184
30 minutes	259	219

5-minute average wind generation for the previous three intervals; and γ_{hour} is the hourly wind forecast value (obtained using the WNPM) for hour where t is in. The coefficients $A - E$ can be tuned to minimize the prediction errors between the actual wind and the forecasted wind power outputs using

$$\min_{P} \sum_{t} (\gamma(P,t) - p_w(t))^2 \qquad (4.2)$$

where P is the coefficient vector $P = \begin{bmatrix} A & B & C & D & E \end{bmatrix}^T$ and $p_w(t)$ is the actual wind generation output at the time t.

The optimal value of the coefficient vector is $P = [0.90.0250.015\,0.01 \quad 0.05]^T$, calculated by using the ERCOT historical data. The performance comparison measured by the mean absolute error (MAE) was given in Table 4.1. By using both the historical information and the forecast from WNPM, the forecast accuracy is significantly improved.

2.2 Method to Determine Regulation Services Requirement

In North America, primary, secondary, and tertiary frequency control are used to maintain system frequency within the desired range. Among them, secondary frequency control is used to maintain the balance between generation (include area interchanges) and load[4] and to provide the frequency support with certain precision dictated by its control performance standards (CPS) score. At ERCOT, regulation up and regulation down services fall into the category of secondary frequency control. Because units need to have very fast ramping capability and there is a significant wear-and-tear effect when providing the RS service, the cost of procuring RSs is usually much higher than that of the energy services. Therefore, it is imperative for ERCOT to determine the appropriate RS requirements for meeting the reliability needs.

---

[4]The balance is achieved by varying outputs of regulation generation through automatic generation control (AGC) every several seconds.

At ERCOT, SCED is running every 5 minutes to dispatch the online generation resources based on the forecasted renewable generation and load. Then, regulation resources are deployed to compensate for the differences between the SCED output (the base points) and the actual system net load to maintain the balance between generation and load. The amount of the RSs needed is determined monthly for the coming month, but the service will be procured at the day-ahead market (DAM) daily. Note that the amount of RS required strongly correlates with the hour of the day and the time of month. Therefore, the general practice is to determine regulation requirement based on real-time or close to real-time generation and load conditions. At ERCOT, the reason that the amount of RS requirement is updated on a monthly basis is to attract more resources to participate in the ancillary service market by eliminating the uncertainties and to hedge the price volatility by purchasing AS at the DAM.

2.2.1 An ERCOT Dispatch Model for Calculating Regulation Requirement

The need for RS strongly depends on how the dispatch process is executed. A simple dispatch model is developed for capturing the essential steps of the ERCOT dispatch procedures. The ERCOT transmission system has expanded quickly to provide necessary transmission accesses for new wind resources and to alleviate the long-term congestion problems [34]. If there is an issue for the deliverability of regulation reserve due to the congestion in the future, the regulation requirement can be studied in smaller regions so that the locational value of regulation service can be accounted for. Therefore, we remove the transmission restrictions in this study.

The base point (BP) for the dispatched generation[5] for the next 5 minutes must meet the instantaneous value of the net load at the beginning of the period because the persistence model is used for the next 5-minute load and wind forecast, which is given by

$$\vec{P}_{BP}^{t} = P_{net-load}^{t-5} = P_{load}^{t-5} - P_{wind}^{t-5} \tag{4.3}$$

where $\vec{P}_{BP}^{t}$ is the base point for time t (in minutes) and $P_{net-load}^{t}$, P_{load}^{t}, and P_{wind}^{t} are the instantaneous net load, load, and wind generation at time t, respectively.

The minute-by-minute imbalance between the load and generation, P_{Δ}^{t}, is

$$P_{\Delta}^{t} = P_{load}^{t} - P_{wind}^{t} - P_{gen}^{t} \tag{4.4}$$

If P_{Δ}^{t} is positive, the regulation up service will be deployed to maintain the frequency for the grid. Otherwise, the regulation down service will be called upon.

---

[5]This is the sum of all base point instructions for generators.

Because the hour-ahead RUC and RTD are deployed to handle the ramping capability of the generation portfolio for large ramp events, it is assumed that the generators can follow the large trend of the net load very well, i.e., when studying the regulation reserve requirement, the dispatched generators follow the base points.[6]

$$P_{gen}^{t} = P_{BP}^{t} \qquad (4.5)$$

Therefore, Eq. (4.4) becomes

$$P_{\Delta}^{t} = P_{load}^{t} - P_{wind}^{t} - P_{BP}^{t} \qquad (4.6)$$

Equation (4.6) represents the minute-by-minute deviation of the net load from the SCED base point (5-minute change in the net load) so it forms the basis for determining the regulation needs.

2.2.2   Regulation Service Requirement

Both deterministic and probabilistic approaches can be used for determining the RS requirement. A deterministic approach is easy to implement. However, meeting the desired reliability for the entire study period is hard because the method may only consider one specific operation condition. Therefore, we proposed a probabilistic-based procedure to determine the RS requirements for the coming month at ERCOT that can (1) take the historical performance into account to meet the pre-determined reliability criterion, (2) consider the variation of wind generation, and (3) validate the grid control performance using a multi-time-scale model.

The calculation for the RS requirement consists of three main steps as described in Table 4.2.

2.2.3   Including Effect of Wind Generation

Wind generation capacity has been added to the ERCOT system at a fast pace in the past few years, and the pace will continue to accelerate. Thus, it is critical to include the effect of the projected wind generation growth when determining the RS requirement. The method used by ERCOT is an improvement to the work in [22] by using the historical chronicle load and wind time-series data since it can preserve the temporal correlation between wind and load. Compared to the previous work using the simulated wind data [22], the proposed method provides a more accurate description of linear dependence between the wind expansion and the RS increment.

As discussed in Sect. 2.2.2, the calculation of RS requirement is based on the statistical analysis of historical deployment of the regulation resources and the

---

[6]In real time, the base point instruction is modified by multiplying a heuristically determined factor with the expected change of the net load. This detail is not modeled here.

Table 4.2 Procedures to calculate RS requirement

Step 1	Calculate the 98.8th percentile of the 5-minute net load changes for the same month of the previous 2 years hour by hour. The data in the same month of the previous 2 years is used with an intention to preserve the seasonal patterns of load changes and wind generation variations. Next, calculate the 98.8th percentile of the regulation up and down services deployed for the same month of the previous 2 years hour by hour. Then, the results will be used to calculate the amount of RS required for each hour to provide an adequate RS 98.8% of the time. The 98.8th percentile is chosen to exclude the extreme values or the outliers. Based on this design, the probability for the regulation resources to be exhausted is 1.2%. In those rare cases when the regulation up resources are exhausted, the contingency reserve will be deployed. Note that if the regulation down resources are exhausted, the wind resources can provide governor actions by ramping down its generation
Step 2	If the average control performance standard 1 (CPS1) score was less than 140% in an hour in the period of the 30 days prior to the time of the study, additional regulation up and down services will be procured for the hour
Step 3	In each month, a backcast of the last month's actual exhaustion rate is performed. An exhaustion index, $\widehat{\zeta}$, calculated every 5 minutes is introduced. If the regulation resource has been fully deployed for the time interval, let $\widehat{\zeta} = 1$. Otherwise, $\widehat{\zeta}$ equals to 0. The exhaustion rate for the nth hour is the exhaustion index averaged for the same hours in the study month, which is given as $$\mu^n = \sum_{i=1}^{N} \sum_{m=1}^{12} \widehat{\zeta}_i^m / (12 \cdot N) \quad (4.7)$$ Where N is the total number of the hours, i is the index of the hour, and m is the index of 5-minute time interval in an hour. If the historical exhaustion rate exceeded 1.2% in any given hour, ERCOT will increase the amount of RS for the coming month

5-minute variations of the net load. To account for the effect of newly installed wind capacity, the following three steps are proposed. A series of scenarios were first built using the historical minute-by-minute wind and load data from different months or different years. Then, the statistical analysis (98.8th percentile) of the 5-minute net load change for the different scenarios, which is the most relevant for determining the RS requirements, was performed and used as the basis for deciding regulation need. The incremental regulation requirement, Ω_j, can be calculated from the historical data for the jth hour by

$$\Omega_j = \frac{r_i^a - r_i^b}{C^a - C^b} \quad (4.8)$$

where r is the regulation required, subscript i indicates the type of the service (1, regulation up; 2, regulation down), superscripts a and b represent the scenario number, and C is the wind installation capacity.

Once the RS need based on the analysis of the historical wind installed capacity, P_{his}^{reg}, is calculated (see Table 4.1), the future RS requirement for the projected wind capacity, P_{future}^{reg}, can be adjusted as

$$P_{future}^{reg} = P_{his}^{reg} + \Omega \cdot (C_{future} - C_{his}) \tag{4.9}$$

where C_{future} and C_{his} are the future and the history wind installed capacity, respectively.

The second term on the right hand of (4.9) represents the change in RS due to the addition of wind generation. The historical ERCOT data has validated the linearity between the added wind capacity and the increase in RS requirement, Ω_j, derived by (4.8), as more details are shown in Sect. 2.3. Note that this linear dependency is assumed in (4.9) for the future years so that one can consider the incremental impact of future wind expansion over the future RS requirement. This linearity for the future years is mainly based on the fact that the 5-minute variations of wind generations are weakly correlated between wind farms or not correlated with those of the load in large geographic areas such as ERCOT. The value of this linear dependency, Ω_j, can be also influenced by the control settings and the CPS1 compliance requirements.

2.2.4 Adequacy of Regulation Reserves

To evaluate regulation reserve adequacy at ERCOT, a multi-time-scale model is developed to simulate grid operation at a resolution of 1 second for a 24-hour study period. All relevant scheduling and operation processes within the study period can be modeled, including SCED, system inertia, generator governor response, load variations with respect to frequency changes, AGC actions, and deployment of regulation resources, as shown in Fig. 4.2.

The existing ERCOT system model and historical data are used to calibrate the simulation tool before it is applied to the RS adequacy study. The model allows the ERCOT engineers to examine major operational or market changes for assessing the possible impacts on grid operation and reliability.

2.3 Procedures to Determine Regulation Service Requirement

The section presents the results obtained from the case study on quantifying the ERCOT system's RS requirement.

2.3.1 Regulation Service Requirements

Operational data in 2014 are used to evaluate the performance of the proposed methods. Field data shows that the amount of additional RS need changes linearly with additional wind capacity as shown in Fig. 4.3. In addition, the change in RS requirement resulting from additional 1000 MW of wind generation varies with the hour of the day and seasons. In Fig. 4.3, we can see that the regulation up service

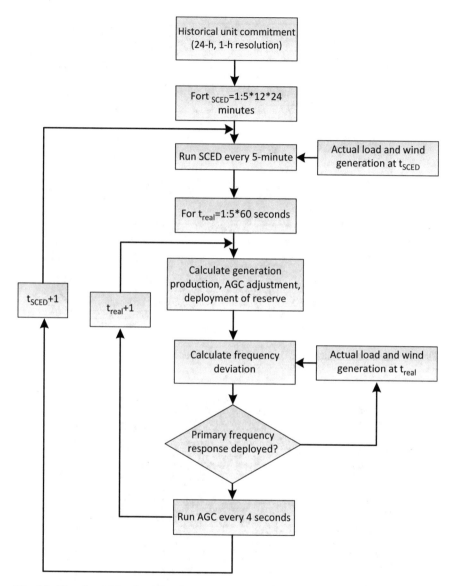

Fig. 4.2 Flowchart of the simulation tool

needs are affected the most in the following hours when adding 1000 MW wind into the ERCOT grid: 19:00–21:00 (January to April), 9:00–14:00 (October and December), 7:00–9:00 (February and October), and 1:00–5:00 (March to June).

Based on the approach described in Sect. 2.2, the amount of regulation down and regulation up services can be calculated each month after considering the effect of wind generation. As shown in Fig. 4.4, the maximum values of the regulation up service requirements occur in the morning throughout the year. This is because the

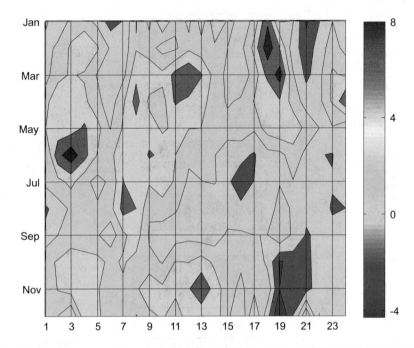

Fig. 4.3 Incremental regulation up requirement in MW per additional 1000 MW wind generation

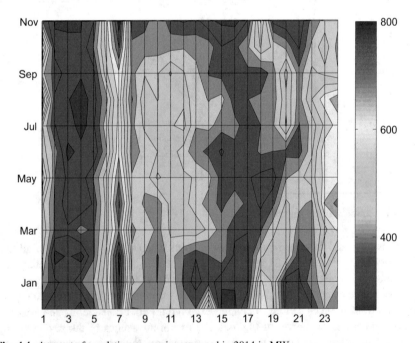

Fig. 4.4 Amount of regulation up service procured in 2014 in MW

morning load ramping-up periods are very consistent throughout the year. Because the effect of wind generation has been considered, a significant portion of the regulation requirement can be contributed to the load. For the regulation down service, which is not shown here, its maximum values occur during the late evening load ramping-down periods.

2.3.2 Evaluation of Exhaustion Rate

The minimum amount of regulation up resources shown in Fig. 4.4 is procured in the DAM in 2014. The exhaustion rate can be calculated backcast month by month for 2014 using (4.7). This index indicates the risk to the grid when there was a shortage in the regulation reserve. The historical monthly exhaustion rate for the regulation up and regulation down service is calculated for 2014, as shown in Fig. 4.5. The annual averaged exhaustion rates for the regulation up and regulation down in 2014 are 1.49% and 0.81%, respectively, both of which are close to the targeted risk level (1.2%). The lower exhaustion rate for the regulation down reserve is partially attributable to the abilities of wind generation units in providing the governor-like actions in response to large frequency excursions.

A bootstrapping method is used to evaluate the confidence intervals for the annual exhaustion risk to account for the uncertainties in both the over-deployment and under-procurement of regulation resources [35]. The probability distribution function (PDF) of the yearly exhaustion rate for 2014 is derived from repeating the resampling process. In this case, 100 runs were executed to reduce the effects of random sampling errors arising from the bootstrap procedure itself. As shown in Fig. 4.6, the exhaustion rates are closely around the mean with a narrow variance. This means that the amount of regulation resources procured meets the targeted

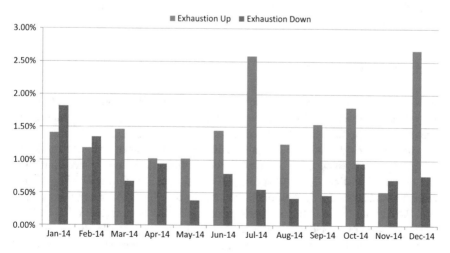

Fig. 4.5 Historical exhaustion rate in 2014

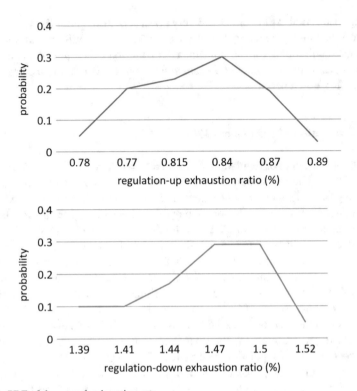

Fig. 4.6 PDF of the annual exhaustion rate

reliability needs with a high confidence level. In addition, the annual average score of CPS1 for ERCOT in 2014 is 163.96%. This further validates the effectiveness of the method used in determining RS requirements in 2014.

2.3.3   Performance Validation

In Sect. 2.2, the new rule was proposed such that more RS should be procured in proportion to the added wind capacity. In this section, the impact of the rule on the system control performance is studied. The following steps describe the application of a simulation model in Fig. 4.2 to ensure the CPS1 compliance given the amount of RS determined in Sect. 2.2.3. First, we validate the performance of the simulation tool described in Sect. 2.2.4 using the historical data. As shown in Fig. 4.7, the simulated and actual system responses are very similar. Then, the simulation tool has been used to study the frequency performance resulting from different RS requirements. One key factor when deciding the RS need is to apply the 98.8th percentile to the 5-minute net load changes. Alternatively, the RS requirement can be calculated

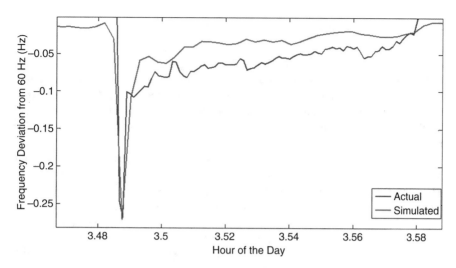

Fig. 4.7   Actual and simulated frequency responses

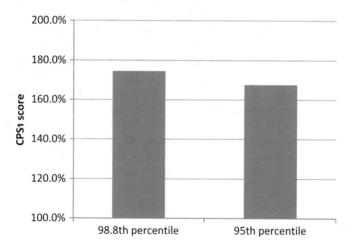

Fig. 4.8   Simulated frequency performance for different regulation up service requirements

based on the 95th percentile of the net load changes. For a typical day (24-hour), these two options have been simulated. As shown in Fig. 4.8, the CPS1 scores are 174.4% and 167.5%, respectively, for the two options. This shows that the RS requirement can be reduced without significantly deteriorating the grid performance by changing the calculation to the 95th percentile of net load changes.

3 Responsive Reserve Requirement

Among three AS products at ERCOT, i.e., regulation up/down, non-spinning reserve, and RRS, RRS resources are used to arrest frequency excursions within a few seconds following generation unit trips. Both the generators providing governor response and load resources with under-frequency relays are eligible to participate in RRS market. The following describes their features in the provision of primary frequency control.

RRS can be provided by online synchronous generators through governor response or governor-like actions to arrest frequency deviations, which is termed as primary frequency response (PFR) service in this chapter. As a single BA, ERCOT must comply with the BAL-003 standard. The frequency response obligation for ERCOT is 413 MW/0.1 Hz. To meet this requirement, ERCOT requires every resource with a speed governor to put the governor in service whenever the resource is online. In addition, the droop setting should not exceed 5%, and the frequency response deadband should not be no more than ± 0.018 Hz.

Load resources (LRs) can also provide RRS if they can be self-deployed to provide a full response within 30 cycles after frequency meets or drops below a certain threshold, which is referred to as fast frequency response (FFR) service. To arrest the quick frequency decline, load resources will be equipped with an under-frequency relay (e.g., triggered if the frequency is dropped under 59.7 Hz). As required by ERCOT, the response time of LRs should be less than 500 ms (including the frequency relay pickup delay and the breaker action time). This makes the response of LRs more effective to mitigate the decline of frequency compared to the generators because a generator needs a few seconds to react to the change in the frequency to provide the primary frequency response. Therefore, the deployment of LRs is able to improve the frequency nadir and is instrumental in preventing frequency from dropping below the involuntary under-frequency load shedding (UFLS) threshold when losing large generation units.

ERCOT must procure a sufficient amount of RRS, including both PFR and FFR, to ensure that the first-stage UFLS is not triggered when losing two largest online generation resources.

3.1 Quantification of PFR and FFR Requirement for Inertias

To decide precisely how much RRS is needed is the prerequisite to the successful market operation where the reliability and economics are balanced. When the RRS requirement is concerned, the key influencing factor is system inertia, which also varies with respect to system loading conditions. In contrast to current practice at other regions where the size of reserve is calculated from a worst scenario, it is desirable to link the RRS requirement with the system inertia at ERCOT which can have a large range. In this way, the impact of anticipated operational conditions (inertia) can be reflected in the RRS quantities.

From the security perspective, the adequate amount of RRS should be allocated to ensure that, for contingencies such as the loss of two largest units, the system frequency should be arrested before triggering UFLS and the frequency nadir should be maintained above 59.3 Hz. In other words, the impact of the loss of two largest generation units can be mitigated efficiently by procuring RRS through a different combination of FFR and PFR.

The following procedures are used to determine the minimum PFR and FFR requirement for different system inertias.

Step 1: Selection of Representative Operation Conditions
The system inertia from all online synchronous generators is calculated as

$$M_{sys} = \sum_{i \in I} H_i \cdot MVA_i \tag{4.10}$$

where H_i and MVA_i are the inertia constant and installation MVA capacity of synchronous machine i, respectively, and I is the online synchronous unit set.

An analysis of annual operation data is conducted to cluster system inertia for ten representative net load levels ranging from 15 GW to 65 GW, as shown in Fig. 4.9. At each net load level, the medium inertia condition is selected as one of the base cases for conducting the dynamic simulation to determine the FRR requirement. Apart from the ten cases, two extreme cases, the maximum and the minimum inertia conditions, are also chosen. Collectively, 12 cases represent the full range of system inertia conditions for determining PFR need.

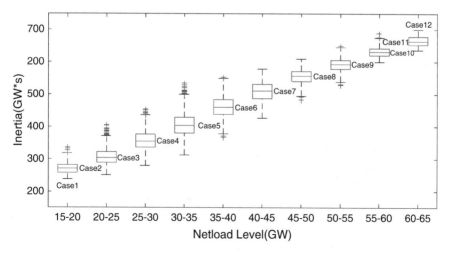

Fig. 4.9 Representative cases for system inertia selection

Step 2: Setup of Dynamic Models

To accurately capture the frequency response of ERCOT system, dynamic models were used with power flow data retrieved from the ERCOT dynamic security assessment databases. These are the same models which are normally used by grid operators to conduct online transient stability analysis for determining the transfer limits across the major transmission lines. The databases include detailed dynamic models of generation units and interruptible load resources. All selected governors are assumed to be operated with 5% droop setting. The triggering frequency of the FFR is set at 59.7 Hz. The pickup delay of the relay is 0.33 seconds, and the breaker action time is 0.083 seconds. Standard ZIP models (constant impedance, constant current, and constant power) are used. The load damping factor is 2%/Hz. Load frequency model is used to represent the sensitivity of constant power and constant current load components to bus frequency. The first stage of the firm load UFLS program is 59.3 Hz, so the under-frequency limit for this study is set at 59.4 Hz to allow a margin of 0.1 Hz. The loss of two largest generation units (2805 MW) is simulated to assess the system frequency response. The frequency response criterion is in compliance with NERC reliability standard, which is given as follows.

"The system frequency shall be arrested between 59.4 Hz and 60.4 Hz so that other generators will not cascade into out-of-step conditions and the firm load UFLS will not be activated."

Step 3: Quantification of Minimum FRR Requirement

In the simulation, the minimum FRR requirement is quantified for a given system inertia condition, and then the same procedure is repeated for different levels of the system inertia. For one of the 12 cases, the system inertia is maintained the same by not changing the commitment status for the generation units. Then, the operation condition is modified by including or removing the governor control or connecting or disconnecting the dynamic models of the interruptible loads. The minimum requirement of FRR for this inertia condition can be identified when the lowest frequency dip equals to 59.4 Hz after two largest units (2805 MW) are taken offline. One example of the frequency response obtained from dynamic simulations is shown in Fig. 4.10. In this case with a system inertia at 354 GW·s, the system frequency is arrested at exactly 59.4 Hz after two largest generation units are tripped. Thus, the minimum FRR requirement is 1240 MW of PFR in addition to 1450 MW of FFR. Through a large number of simulations, a family of the curves can be derived to reveal how the minimum FRR requirements change when the combination of PFR/FRR differs.

Step 4: Derivation of Equivalency Ratio for All Cases

Through the experiments, it is observed that when the amount of FFR increases, the required amount of PFR will be reduced to protect the system integrity. As the relationship between PFR and FFR requirements is approximately linear, the slope of the lines can be viewed as an equivalency ratio, m, between PFR and FFR requirements. For example, when the system inertia condition is at 354 GW·s, if m is 1.4 and the minimum total FRR requirement is 3370 MW, the constraint can be described by

Fig. 4.10 Simulated frequency response when the system inertia is 354 GW·s

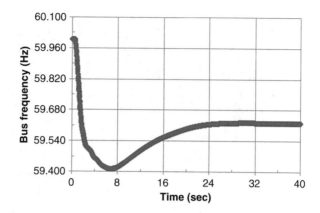

Table 4.3 Minimum FFR requirement and equivalency ratios

Case no.	Inertia (GW·s)	FRR$_{min}$ (MW)	Equivalency ratio (m)
1	239	5200	2.2
2	271	4700	2.0
3	304	3750	1.5
4	354	3370	1.4
5	403	3100	1.3
6	459	3040	1.25
7	511	2640	1.13
8	556	2640	1.08
9	593	2240	1
10	631	2280	1
11	664	2140	1
12	700	2140	1

$$1.4 \cdot P_{FFR} + P_{PFR} \geq 3370 \tag{4.11}$$

It means that 1 MW of FFR is 1.4 times as effective as 1 MW of PFR in arresting the frequency decline. Alternatively, each 1 MW of FFR can be replaced by 1.4 MW of PFR.

Table 4.3 summarizes the results of all 12 cases for the ERCOT grid. In general, FFR is more effective than PFR when arresting the frequency decline at low inertia conditions (e.g., Cases 1 through 6). However, the effectiveness decreases when the system inertia increases. When inertia is higher than 593 GW·s, FFR and PFR become equally effective. This is because there are more generators online under heavy loading conditions so that the overall system inertia is higher, making the rate of change of frequency following the disturbance much less than that in a low inertia condition. As a result, aggregated slow-acting governor-like response has enough time to react to the loss of generation. The results also demonstrate that fast load

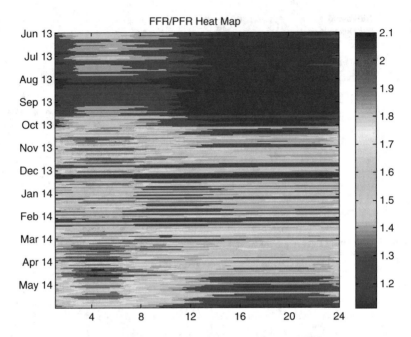

Fig. 4.11 Heat map of historical equivalency ratios between PFR and FFR

response is more valuable in arresting the frequency drops when response speed is more critical, which is the case in low inertia systems.

The historical equivalency ratios for the ERCOT grid in 1 year are then calculated. As shown in Fig. 4.11, in hot summer peak load hours (noon to evening), m equals to 1.0. For early morning hours in shoulder and winter months, m can be as high as Sect. 2.2.

3.2   RRS Requirement

One key factor influencing how much RRS is needed is the grid system inertia. This dependence is considered here when deciding the need for RRS. The basic approach for determining RRS requirement consists of two steps: (1) to project system inertia conditions for the next year by creating a time-series data for 8760 hours and (2) to map from the projected inertia condition to the RRS need. Note that the minimum RRS requirement is determined for six 4-hour blocks every day and the quantities are also the same for each month.

The two-step procedures are described as follows.

The first step is to calculate future inertia conditions for each 4-hour interval using the historical inertia information. The historical inertia is recorded according to (4.10), which reflects the commitment statues of thermal units and their inertia

Table 4.4 Block number for different hour ending

Hour ending	1–2, 23–24	3–6	7–10	11–14	15–18	19–22
Hour block	1	2	3	4	5	6

constant. For each day, 24 hours are placed into six 4-hour blocks—see Table 4.1. For example, the hours whose hour endings are 1–2 or 23–24 are clustered as the first hour block.

The projection of future inertia condition is then based on expected diurnal load and wind patterns for the same hour block of the same month in the past 2 years. In other words, to determine the projected inertia condition for the hour block $j_{\text{hour block}}$ of the month i_{month} for the year k_{year}, the historic system inertia conditions for the same 4-hour block (see Table 4.4) of the same month from two previous years are collected, which is given by the data set.

$$\Phi = \left\{ M_{k_{\text{year}-1}}^{i_{\text{month}}, j_{\text{hour block}}}, M_{k_{\text{year}-2}}^{i_{\text{month}}, j_{\text{hour block}}} \right\} \tag{4.12}$$

where $M_{k_{\text{year}-1}}^{i_{\text{month}}, j_{\text{hour block}}}$ and $M_{k_{\text{year}-2}}^{i_{\text{month}}, j_{\text{hour block}}}$ are the system inertia of the hour block $j_{\text{hour block}}$ of the month i_{month} from the year $k_{\text{year}-1}$ and $k_{\text{year}-2}$, respectively.

A percentile is then applied to this data set, Φ, to produce a future inertia condition for each 4-hour block of the same month for the next year, $M_{k_{\text{year}}}^{i_{\text{month}}, j_{\text{hour block}}}$. A 70th percentile is chosen to avoid the over-procurement of RRS while covering the reliability risk of low inertia operation hours.[7]

The second step is to determine the minimum amount of RRS requirement based on the projected future inertia conditions, $M_{k_{\text{year}}}^{i_{\text{month}}, j_{\text{hour block}}}$, through a look-up table. This look-up table, which is produced from a series of dynamics studies, constitutes a mapping from an inertia level to the corresponding RRS need. The procedures on how to create this table will be given in more details in the next section. As required by the current operational procedures, the FFR resources should restore their responsibility within 3 hours once they are deployed and recalled. Thus, during this 3-hour restoration time, the grid reliability may be at risk if the exhausted FFR resources are not available for deployment for another disturbance. Due to this reason, ERCOT only allows FFR resources to contribute to up to 60% of the total RRS need. When calculating the RRS requirement, it is assumed that the share of FFR in the provision of RRS is limited to 60%. These RRS amount will then be published in six separate blocks covering 4-hour intervals by month.

Figure 4.12 depicts the RRS requirement for the year of 2018. The highest RRS requirements occur at the midnight hours during the shoulder months, while the lowest RRS requirements coincide with the summer load peak hours.

---

[7] The P-th percentile of a list of N ordered values (sorted from least to greatest) is the largest value in the list such that at least P percent of the data is greater than or equal to that value. This is obtained by first calculating the ordinal rank and then taking the value from the ordered list that corresponds to that rank.

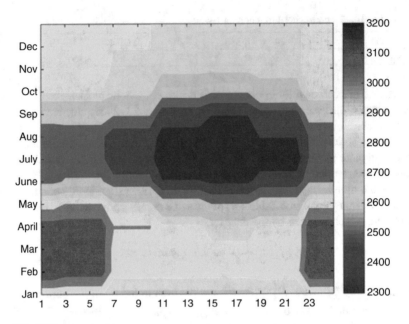

Fig. 4.12 RRS procurement

4 Non-spinning Reserve

4.1 Background

Non-spinning reserve (Non-Spin) consists of generation resources capable of being ramped to a specified output level within 30 minutes or load resources that are capable of being interrupted within 30 minutes and that are capable of running (or being interrupted) at a specified output level for at least 1 hour. Non-Spin may be deployed to replace loss of generating capacity, to compensate for load/wind/solar forecast and/or forecast uncertainty on days in which large amounts of reserve are not available online, to address the risk of net load ramp, or when there is a limited amount of capacity available for SCED.

The periods when load is increasing and wind/solar is decreasing require other generation resources to increase output or come online quickly to compensate for the sudden net load increases. As a result, net load ramp risk should be accounted for in the determination of Non-Spin requirements. While net load forecast analysis may cover reserves required for forecast uncertainty, it may not necessarily cover exposure to the loss of generation and net load ramp risk. Due to this risk, it may be necessary for ERCOT to have additional reserves available to protect against forecast uncertainty and forced outages of thermal resources within an operating day.

Examples of circumstances when Non-Spin has been used are:

- Across peak hours during spring and fall months when hotter than expected weather with large amounts of capacity offline resulted in energy emergency alert (EEA) events.
- Afternoons during summer seasons when high loads and unit outages outstripped the capability of base load and normal cyclic units.
- Cold weather events when early morning load pickup outpaced the ability of generation to follow.
- Major unit trips when large amounts of spinning reserve were not online.
- During periods when the net load (load-wind-solar) increased more than forecasted.

There are four situations that will cause Non-Spin to be deployed:

- Detection of insufficient capacity for energy dispatch during periodic checking of available capacity.
- Disturbance conditions such as a unit trip, sustained frequency decay, or sustained low-frequency operations.
- SCED not having enough energy available to execute successfully.
- When offline generation resource providing Non-Spin is the only reasonable option available to the operator for resolving local issues.

In each of these cases, the ERCOT operator will make the final decision and initiate the deployment. The ERCOT operator shall deploy Non-Spin in amounts sufficient to respond to the operational circumstances. This means that Non-Spin may be deployed partially over time or may be deployed in its entirety. If Non-Spin is deployed partially, it shall be deployed in increments of 100% of each resource's capacity. To support partial deployment, for each hour of the operating day, ERCOT shall, following the day-ahead market (DAM), rank the resources supplying Non-Spin in an economic order based on DAM settlement point prices. Partial Non-Spin deployment and recall decisions shall be based on each resource's economic cost order.

More specifically, ERCOT may deploy Non-Spin, which has not been deployed as part of a standing online Non-Spin deployment, under the following conditions:

- When (HASL[8] – Gen – Intermittent Renewable Resource (IRR) Curtailment) – (30-minute net load ramp) <0 MW, deploy half of the available Non-Spin capacity.
- When (HASL – Gen – IRR Curtailment) – (30-minute net load ramp) <-300 MW, deploy all of the available Non-Spin capacity.
- When Physical Responsive Capability (PRC) <3200 MW and not expected to recover within 30 minutes without deploying reserves, deploy all or a portion of the available Non-Spin capacity.
- When PRC <2500 MW, deploy all of the available Non-Spin capacity.

---

[8]HASL: high ancillary service limit.

- When the North-to-Houston Voltage Stability Limit Reliability Margin <300 MW, deploy Non-Spin (all or partial) in the Houston area as needed to restore reliability margin.
- When offline generation resources providing Non-Spin are the only reasonable option available to the operator for resolving local issues, deploy available Non-Spin capacity on only the necessary individual resources.

If a condition other than those listed above indicates that additional capacity may need to be brought online resources to manage reliability, operators will evaluate the system condition and deploy Non-Spin as needed if no other better options are available to resolve the system condition. Under emergency, the emergency process will govern the deployment of Non-Spin.

4.2   Net Load Forecast Error (NLFE) Analysis

The historical net load forecast error (NLFE) is taken as the primary basis to determine Non-Spin requirement. At ERCOT, the forecast for load and wind/solar generation are produced separately. NLFE can be obtained as follows:

$$\varepsilon_{nl} = \varepsilon_l - \varepsilon_w - \varepsilon_s \qquad (4.13)$$

where ε_{nl} is the net load forecast error, ε_l is the load forecast error, ε_w is the wind forecast error, and ε_s is the solar forecast error.

An analysis was performed to understand the variations and trends in NLFE, using the 3-hour-ahead and 6-hour-ahead forecast and actual data for 1 year. The probability distribution function (PDF) of the NLFE gives the relative likelihood of NLFE to take on any given value. The probability of the NLFE falling within a certain range is given by the integral of the NLFE's density over that particular range.

The distribution of NLFE using the 3-hour-ahead forecast has a higher kurtosis than the distribution using the 6-hour-ahead forecast, as shown in Fig. 4.13. This is mainly because the 3-hour-ahead forecast is more accurate than the 6-hour-ahead forecast resulting in lower values of NLFE. Overall, in either case, the distributions are skewed toward the under-forecast side, i.e., NLFE >0. Also, it can be seen that the distributions are non-Gaussian. Since the distributions are not "normal," standard deviation cannot be used to evaluate the confidence intervals. To compare the values of NLFE at different confidence intervals, cumulative distribution function (CDF) was used.

The CDF of NLFE can be used to compare the values of NLFE at different confidence intervals. Currently, 85th–95th percentile is used as the confidence interval for NLFE for calculating the Non-Spin requirement. The curves in Fig. 4.14 show a larger variation in NLFE in the lower and the higher confidence interval pertaining to longer tails in the PDF. Specifically, as we move higher up

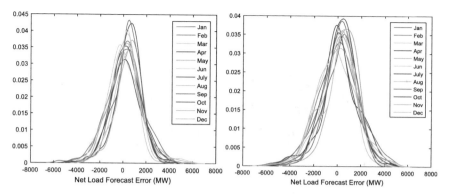

Fig. 4.13 PDF of monthly NLFE (left, 3-hour-ahead NLFE; right, 6-hour-ahead NLFE)

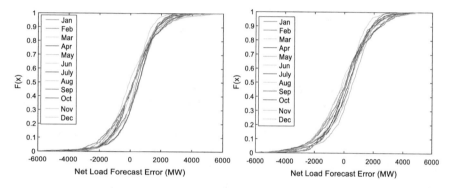

Fig. 4.14 CDF of monthly NLFE (left, 3-hour-ahead NLFE; right, 6-hour-ahead NLFE)

from the currently used 95th percentile, the NLFE increases significantly than if we were to go lower than the 95th percentile. Considering a higher confidence interval means including more outliers and, conversely, considering a lower confidence eliminates those outliers.

A scatter plot between load forecast error and wind forecast error gives a good idea about the relationship between the load, wind, and net load errors. Each of the points in the plot represents the wind and load over forecast error values for hours as shown in Fig. 4.15. The sign and magnitude of the load and wind forecast error determine whether the net load is being under- or over-forecasted. The top triangular region bordered by the axes and the green line represents the region where the combination of load forecast error and wind forecast error results in net load under-forecast error. The bottom triangular region represents the region where the combination of load forecast error and wind forecast error results in net load over-forecast error. The green line in the middle represents the set of points at which the load forecast error and wind forecast error cancel each other out to produce a zero NLFE. The colored lines in the upper triangular region represent the range of confidence intervals for the NLFE. The area between each of the three colored lines and the

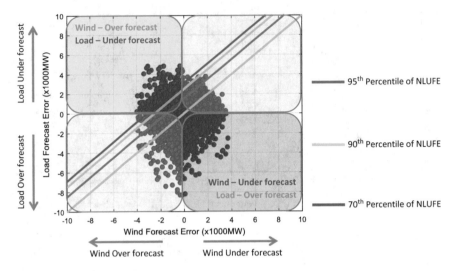

Fig. 4.15 Wind vs. load forecast error (net load under-forecast error: NLUFE)

green line represents the set of data points within that particular confidence interval. Any data point lying outside the range is an outlier and is not considered for further analysis.

The plot can be further divided into four distinct quadrants based on the sign of load and wind forecast errors. The red region on the top left represents the region with the highest reliability threat. In this region, the load is under-forecasted and the wind is over-forecasted. The net effect of combining under-forecast of load and over-forecast of wind generation results in a large, positive value for NLFE. The blue region, on the other hand, represents the region with the least reliability concern. In this region, the actual demand is lower than the forecasted demand. Although this still requires some corrective actions to maintain system frequency close to 60 Hz, this can be efficiently managed using regulation down reserve, governor action, or curtailment of wind or solar generation.

From a reliability perspective, the under-forecast side possesses a higher reliability threat than the over-forecast error when the actual net load is greater than the forecasted net load. Since Non-Spin is deployed to cover for the gap between the generation capacity and demand so as to maintain surplus reserves, the net load over-forecast data should not be used when calculating the Non-Spin requirement.

4.3   Net Load Ramp-Up

A sudden rise in the magnitude of net load over a short period of time is considered a high reliability risk as the deficient amount of generation has to be quickly dispatched to meet the system demand. Small fluctuations in load are usually covered by regulation reserve services, provided by online resources, which are running every 4 seconds to compensate for the mismatch between generation and demand. However, if such an increase in net load is substantial and over a period of a few hours, more generation has to be committed dispatched to meet the increasing demand. However, these hours, when the net load is ramping up, have a high reliability risk as a failure to dispatch the adequate amount of generation, due to either forecast error or forced outages, will result in deficient generation. At this time, sufficient Non-Spin has to be available to cover for any outages or forecast errors.

During the summer months, the peak loads occur during the afternoon when the wind generation is low. The net load ramps up during the morning hours when the load is ramping up and the wind is ramping down.

During the winter months, the peak loads occur during the morning and the evening hours. The wind generation is relatively high during winter and spring, but there is a down ramp during the early morning hours when the load is ramping up.

Each hour for every month in a year can be associated with a net load ramp-up risk factor, which gives the relative risk of occurrence of a net load ramp-up with respect to the seasonal peak value. The net load ramp-up risk factor is calculated by normalizing hourly net load ramp-up over the seasonal net load ramp-up peak value as follows:

$$\delta_{\text{NLRU}} = \frac{\gamma_{h-\text{NLRU}}}{\gamma_{p-\text{NLRU}}} \tag{4.14}$$

where δ_{NLRU} is the net load ramp-up risk factor, $\gamma_{h-\text{NLRU}}$ is the hourly net load ramp-up, and $\gamma_{p-\text{NLRU}}$ is the seasonal net load ramp-up peak.

The values of net load ramp-up risk factor vary from 0 to 1. Since the Non-Spin requirement is calculated in 4-hour blocks for each month's 24-hour period, the similar calculation is also applied to the net load ramp-up risk factor which is averaged for each of six 4-hour blocks per month.

To improve the efficiency of procurement of Non-Spin, the amount of Non-Spin procured for a particular hour should reflect the risk of net load ramp-up events. To do this, the risk factor can be taken as a basis for determining the confidence interval chosen for the NLFE for a particular hour. The CDF plot from the previous section shows that reducing the confidence interval will result in a lower value of NLFE and a higher confidence interval will result in a higher value of NLFE. Hence, if the risk factor were to be associated with a particular confidence interval such that the resulting value of NLFE reflected the risk of net load ramp-up, this would result in a more efficient way of procuring Non-Spin.

A linear function is used to map the 4-hour averaged net load ramp-up risk factor to the confidence interval for NLFE to be considered for the Non-Spin calculation. A minimum of 85th percentile and a maximum of 95th percentile confidence interval are chosen for NLFE. Periods where the risk of net load ramp is highest will use 95th percentile and 85th percentile for periods with lowest risks.

4.4 Adjustment to Non-Spin Need by Considering Forced Outage

Non-Spin can be deployed to cover the reliability risk caused by a forced outage, a forecast error, or a combination. The forced outage will make a less amount of the generation capacity available. For a normal day, operators can commit other offline resources to substitute the lost generation capacity. However, for a supply tight day when all of available generation resources are already online, there is no extra capacity except Non-Spin. The current methodology adds an intra-day forced outage table to the confidence interval of NLFE previously calculated. The proposed intra-day forced outage table has been built using 75th percentile of the maximum accumulated forced outage in a 6-hour block for each month in the previous 3 years. For the purposes of building this table (see Fig. 4.16), a conservative 75th percentile coverage is recommended due to the observation that the extreme forced outages tend to not occur at the same time as the highest NLFE.

ERCOT has seen significant growth in installed wind and solar capacity from one year to the next year; an increase in wind and solar capacity also tends to increase the quantity of error in their respective forecasts. Hence, ERCOT's reliance on historical wind and solar forecast errors alone creates a possibility of under-estimation of the Non-Spin requirement. To address this, ERCOT will include the impact of increase in over-forecast error from the expected growth in wind and solar generation installed capacity into the future Non-Spin requirement. The net wind impact is

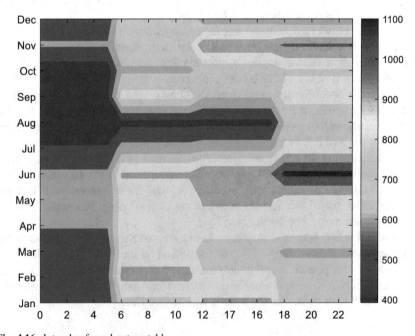

Fig. 4.16 Intra-day forced outage table

calculated by a multiplication of the projected wind capacity growth between the same month of the current year and the next year and incremental adjustment to Non-Spin value per 1000 MW of incremental wind generation capacity. The incremental wind adjustment to the Non-Spin value per 1000 MW increase in wind installed capacity is calculated as the change in the 50th percentile of the historical wind over-forecast error for 4-hour blocks of each month in the past 5 years, which is then normalized to per 1000 MW of installed wind capacity. In a similar way, the net solar impact is calculated by a multiplication of the projected solar capacity growth between the same month of the current year and the next year and incremental adjustment to Non-Spin value per 1000 MW of incremental solar generation capacity. The incremental solar adjustment to the Non-Spin value per 1000 MW increase in solar installed capacity is calculated as the change in 50th percentile of the historical solar over-forecast error for 4-hour blocks of each month in the past 3 years, which is then normalized to per 1000 MW of installed solar capacity. Figures 4.17 and 4.18 reflect the additional Non-Spin adjustments per 1000 MW of installed wind and solar capacity.

4.5 Procedures to Determine Non-Spin Need

Analyses for Non-Spin requirements are conducted using data from the same month of the previous 3 years. For the purpose of determining the amount of Non-Spin to purchase for each hour of the day, hours will be placed into 4-hour blocks. The procedures to determine Non-Spin need can be summarized as follows.

Step 1: The net load uncertainty for the analyzed days for all hours which are considered to be part of a 4-hour block will be calculated, and a percentile will be assigned to this block of hours based on the risk of net load ramp. The same calculation will be done separately for each block. Net load is defined as the ERCOT load minus the estimated uncurtailed total output from intermittent renewable resource (IRR), which includes both wind-powered generation resources (WGRs) and photovoltaic generation resources (PVGR) at a point in time. The wind and solar forecast and the mid-term load forecast used are the updated values as of 6 hours prior to each operating hour. The net load uncertainty is then defined as the difference between the highest 5-minute net load within the hour and the forecasted net load.

Step 2: The risk of net load ramp is determined based on the change in net load over an hour divided by the highest observed net load for the season. The fixed value of percentile ranging between 85th percentile and 95th percentile will be assigned to the net load forecast uncertainty calculated previously. ERCOT then determines the Non-Spin requirement using the 85th to 95th percentile of hourly net load uncertainty from the same month of the previous 3 years.

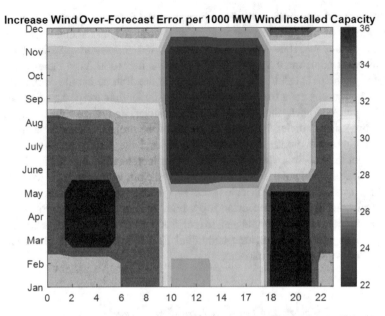

Fig. 4.17 Wind over-forecast error adjustment table (wind over-forecast error adjustment table tracks estimated increase in wind over-forecast error per 1000 MW increase in installed wind capacity)

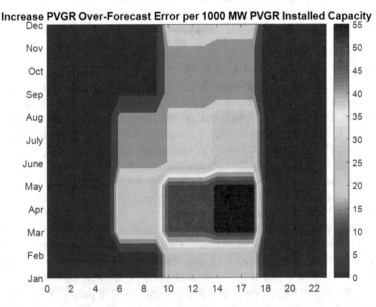

Fig. 4.18 Solar over-forecast error adjustment table (solar over-forecast error adjustment table tracks estimated increase in solar over-forecast error per 1000 MW increase in installed solar capacity)

Step 3: The Non-Spin requirement for the month for each block is calculated using the assigned percentile (based on the risk of net load ramp) for the block minus the average reg-up requirement during the same block of hours ("Non-Spin block").

Step 4: ERCOT will include the impact of increase in over-forecast error from the expected growth in wind and solar generation installed capacity into the future Non-Spin requirement. The net wind impact is calculated by a multiplication of the projected wind capacity growth between the same month of the current year and the next year and incremental MW adjustment to Non-Spin value per 1000 MW of incremental wind generation capacity. A similar procedure is also applied to solar generation.

Step 5: The Non-Spin requirement for each hour in the month is calculated by adding an adjustment that accounts for intra-day forced outage of thermal resources to the previously calculated "Non-Spin block" quantity that the hour falls in. This forced outage adjustment is calculated as the 75th percentile of the historical intra-day forced outages (accumulated since midnight) for 6-hour blocks of each month in the past 3 years.

ERCOT will purchase Non-Spin such that the combination of Non-Spin and reg-up services covers the uncertainties of net load forecast errors depending on the net load ramp risk and intra-day forced outages.

Figure 4.19 shows the minimum Non-Spin requirement in 2022. It can be observed that a higher amount of Non-Spin is needed for morning or evening peak load hours, while less Non-Spin is procured during midnight hours.

5 Summary

The primary ancillary services at ERCOT include regulation up, regulation down, responsive reserves, and non-spinning reserves. Market participants may self-schedule ancillary services or have them purchased on their behalf by ERCOT.

ERCOT procures responsive reserves to ensure that the system frequency can quickly be restored to appropriate levels after a sudden, unplanned outage of generation capacity. Non-spinning reserves are provided from offline resources that can start quickly to respond to contingencies and to restore responsive reserve capacity. Regulation reserves are capacity that responds every 4 seconds, either increasing or decreasing as necessary to keep output and load in balance from moment to moment. The quantity of regulation needed is affected by the accuracy of the supply and demand reflected in the 5-minute dispatch.

This chapter describes in depth the methodology used at ERCOT to determine ancillary service requirement, which is aligned with how ancillary services are used and also considers the incremental impact of wind and solar generation resources. The average total ancillary services requirement in 2020 was about 4800 MW, although the quantity of reserves held varies hour to hour. The average ancillary

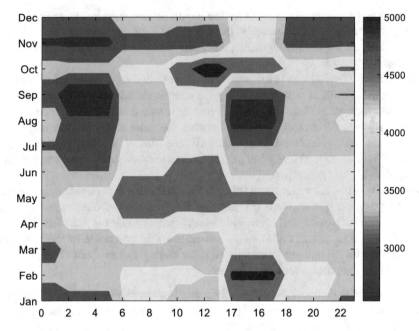

Fig. 4.19 Non-spin requirement

Table 4.5 Ancillary service price

	Responsive reserves	Non-spinning reserves	Regulation up	Regulation down
Ancillary service price ($/MWh)	$11.40	$4.45	$11.32	$8.45

service cost per MWh of load was $1.00 per MWh in 2020. Table 4.5 lists the hourly averaged ancillary service price in 2020.

As more intermittent, inertia-less wind and solar resources are being connected to the ERCOT grid, the system reliability need is also evolving. This drives both the increasing need for new ancillary services and the increase in the quantities for the existing ancillary products. ERCOT continues to tune the ancillary service method- ology to ensure that the system reliability need is met.

References

1. Warren Katzenstein and James Hansell, Adding Appropriate Noise to Interpolation of Hourly Profiles, KEMA/DNV, 2013.

2. J. Charles Smith, Michael R. Milligan, Edgar A. DeMeo, and Brian Parsons, "Utility wind integration and operating impact state of the art," IEEE Transactions on Power Systems, 2007, 22, (3), pp. 900–908.
3. Yuri V. Makarov, Clyde Loutan, Jian Ma, and Phillip de Mello, "Operational impacts of wind generation on California power systems," IEEE Transactions on Power Systems, 2009, 24, (2), pp. 1039–1050.
4. Yuri Makarov, Pengwei Du, Michael Kintner-Meyer, Chunlian Jin, and Howard Illian, "Sizing energy storage to accommodate high penetration of variable energy resources," IEEE Transactions on Sustainable Energy, 2012, 3, (1), pp. 34–40.
5. Ross Baldick, "Wind and energy markets: a case study of Texas," IEEE Systems Journal, 2012, 6, (1), pp. 27–34.
6. S.-H. Huang, D. Maggio, K. McIntyre, V. Betanabhatla, J. Dumas and J. Adams, "Impact of wind generation on system operations in the deregulated environment: ERCOT experience," Proceedings of IEEE PES General Meeting, Calgary, CA, Intersociety of Electrical and Electronics Engineers, Piscataway, NJ, June 2009.
7. Yann G. Rebours, Daniel S. Kirschen, Marc Trotignon, and Sébastien Rossignol, "A survey of frequency and voltage control ancillary services—Part I: technical features," IEEE Transactions on Power Systems, 2007, 22, (1), pp. 350–357.
8. Yann G. Rebours, Daniel S. Kirschen, Marc Trotignon, and Sébastien Rossignol, "A survey of frequency and voltage control ancillary services—Part II: economic features," IEEE Transactions on Power Systems, 2007, 22, (1), pp. 358–366.
9. Erik Ela, Michael Milligan, and Brendan Kirby, "Operating reserves and variable generation," Technical ReporNREL/TP-5500, National Renewable Energy Laboratory, Golden, Colorado, August 2011.
10. Erik Ela, Brendan Kirby, Nivad Navid, J. Charles Smith, "Effective ancillary services market designs on high wind power penetration systems," IEEE Power and Energy Society General Meeting, San Diego, California, July 22–26, Intersociety of Electrical and Electronics Engineers, Piscataway, NJ, 2012, pp. 1–7.
11. Hannele Holttinen, Michael Milligan, Erik Ela, Nickie Menemenlis, Jan Dobschinski, Barry Rawn, Ricardo J. Bessa, Damian Flynn, Emilio Gómez-Lázaro, and Nina K. Detlefsen, "Methodologies to determine operating reserves due to increased wind power," IEEE Transactions on Sustainable Energy, 2012, 3, (4), pp. 713–723.
12. Ronan Doherty and Mark O'Malley, "A new approach to quantify reserve demand in systems with significant installed wind capacity," IEEE Transactions on Power Systems, 2005, 20, (2), pp. 587–595.
13. Juan M. Morales, Antonio J. Conejo, and Juan Pérez-Ruiz, "Economic valuation of reserves in power systems with high penetration of wind power," IEEE Transactions on Power Systems, 2009, 24, (2), pp. 900–910.
14. Miguel A. Ortega-Vazquez, and Daniel S. Kirschen, "Estimating the spinning reserve requirements in systems with significant wind power generation penetration," IEEE Transactions on Power Systems, 2009, 24, (1), pp. 114–125.
15. François Bouffard, Francisco D. Galiana, and Antonio J. Conejo, "Market-clearing with stochastic security—Part I: formulation," IEEE Transactions on Power Systems, 2005, 20, (4), pp. 1818–1826.
16. Miguel A. Ortega-Vazquez, and Daniel S. Kirschen, "Optimizing the spinning reserve requirements using a cost/benefit analysis," IEEE Transactions on Power Systems, 2007, 22, (1), pp. 24–33.
17. Ricardo J. Bessa, Manuel A. Matos, Ivo C. Costa, Leonardo Bremermann, Ivan Gustavo Franchin, Rui Pestana, Nélio Machado, Hans-Peter Waldl, and Christian Wichmann, "Reserve setting and steady-state security assessment using wind power uncertainty forecast: a case study," IEEE Transactions on Sustainable Energy, 2012, 3, (4), pp. 827–836.
18. Manuel A. Matos and R. J. Bessa, "Setting the operating reserve using probabilistic wind power forecasts," IEEE Transactions on Power Systems, 2011, 26, (2), pp. 594–603.

19. M. Milligan, P. Donohoo, D. Lew, E. Ela, B. Kirby, H. Holttinen, E. Lannoye, D. Flynn, M. O'Malley, N. Miller, and P.B. Eriksen, "Operating reserves and wind power integration: an international comparison," In proc. 9th International Workshop on large-scale integration of wind power into power systems, October 2010, pp. 18–29.
20. Chavez, Hector, Ross Baldick, and Sandip Sharma, "Regulation adequacy analysis under high wind penetration scenarios in ERCOT nodal," IEEE Transactions on Sustainable Energy, 2012, 3(4), pp. 743–750.
21. D. Lew, G. Brinkman, E. Ibanez, B.M. Hodge, J. King, "Western wind and solar integration study phase 2," National Renewable Energy Laboratory, Golden, Colorado, 2013.
22. General Electric, "Analysis of wind generation impact on ERCOT ancillary services requirements", Schenectady, NY, March 28, 2008. http://www.uwig.org/attchb-ercot_a-s_study_final_report.pdf?bcsi-ac-a7ccf25998285488=240D6A9C000000029Ms3Um5elR4RfHNUn8SlEZ4/7vSaBwAAAgAAAI03HQCEAwAAAgAAABevAAA=.
23. Nivad Navid, Gary Rosenwald, Ramp Capability Product Design for MISO Markets, Midcontinent Independent System Operator, July 2013.
24. "MMS Look-Ahead SCED: Phase 1 Requirements," ERCOT, 2012.
25. Y. Gu, and L. Xie, "Early detection and optimal corrective measures of power system insecurity in enhanced look-ahead dispatch," IEEE Transactions on Power Systems, 2013. 28, (2), pp. 1297–1307.
26. "ERCOT Nodal Protocol," ERCOT, Taylor, TX, 2015.
27. Tengshun Peng, Dhiman Chatterjee, Pricing Mechanism for Time Coupled Multi-interval Real-Time Dispatch, 6/24/2013, MISO http://www.ferc.gov/CalendarFiles/20140411125224-M2%20-%20Peng.pdf
28. F. Gao, A. Hallam, and C-N Yu, "Wind generation scheduling with pump storage unit by collocation method," Proceedings of IEEE PES General Meeting, Calgary, CA, Intersociety of Electrical and Electronic Engineers, Piscataway, NJ, June 2009, pp. 1–7.
29. H. Hui, C.-N. Yu and S. Moorty, "Reliability unit commitment in the new ERCOT nodal electricity market," Proceedings of IEEE PES General Meeting, Calgary, CA, Intersociety of Electrical and Electronics Engineers, Piscataway, NJ, June 2009, pp. 1–7.
30. Hailong Hui, Chien-Ning Yu, Resmi Surendran, Feng Gao and Sainath Moorty, "Wind generation scheduling and coordination in ERCOT Nodal market," 2012 IEEE Power and Energy Society General Meeting, 2012, Intersociety of Electrical and Electronics Engineers, Piscataway, NJ, pp.1–8, pp. 1–7.
31. J. Wang, A. Botterud, R. Bessa, H. Keko, L. Carvalho, D. Issicaba, J. Sumaili, V. Miranda, "Wind power forecasting uncertainty and unit commitment," Applied Energy, 2011, 88, (11), pp. 4014–4023.
32. E. M. Constantinescu, V. M. Zavala, M. Rocklin, S. Lee, M. Anitescu, "Unit commitment with wind power generation: integrating wind forecast uncertainty and stochastic programming," Argonne National Laboratory (ANL), Argonne, IL, 2009.
33. C. Monteiro, R. Bessa, V. Miranda, A. Botterud, J. Wang, and G. Conzelmann, "Wind power forecasting: state-of-the-art 2009," ANL/DIS-10-1, Argonne National Laboratory, Argonne, IL November 2009. (see also http://www.dis.anl.gov/projects/windpowerforecasting.html).
34. W. Lasher, "Transmission planning in the ERCOT interconnection," ERCOT, Taylor, TX, 2011 [Online]: http://www.ercot.com/content/news/presentations/2011/ERCOT%20Transmission%20Planning%20DOE%20EAC%2010-20-11.pdf
35. B. Efron, R. Tibshirani, *An Introduction to the Bootstrap*, 1993, Chapman & Hall/CRC.

Chapter 5
Design of New Primary Frequency Control Market for Hosting Frequency Response Reserve Offers from Both Generators and Loads

1 Introduction of Frequency Control

Maintaining system frequency at its target value is critical to power grid operation [1–4]. The overall task of controlling frequency is organized in three levels in North America, namely, the primary, secondary, and tertiary frequency control.

The primary frequency control is actions provided by the interconnection to arrest and stabilize frequency in response to frequency deviations. Primary control comes from automatic generator governor response, load response (typically from motors), and other devices that provide an immediate response based on local (device-level) control systems. Primary frequency response is designed for severe contingencies such as large generators tripping offline. It involves an automatic and quick action of a generator to change its output in response to a large frequency deviation (within seconds).

Secondary frequency control is the actions provided by an individual balancing authority or its reserve sharing group to correct the resource-load unbalance that created the original frequency deviation, which will restore both scheduled frequency and primary frequency response. Secondary control comes from either manual or automated dispatch from a centralized control system. Secondary frequency control is designed for small frequency deviations and delivers a slow response. It is activated by automatic generation control (AGC) to provide an automatic response. It is able to maintain system frequency within the boundary when there is a small demand change.

Tertiary frequency control is the actions provided by balancing authorities on a balanced basis that are coordinated so there is a net-zero effect on area control error (ACE). Examples of tertiary control include dispatching generation to serve native load, economic dispatch, dispatching generation to affect interchange, and re-dispatching generation. Tertiary control actions are intended to replace secondary control response by reconfiguring reserves. Tertiary control operates on a longer time scale (minutes) than primary and secondary frequency control. Its main purpose

P. Du, *Renewable Energy Integration for Bulk Power Systems*, Power Electronics and Power Systems, https://doi.org/10.1007/978-3-031-28639-1_5

is to replenish reserves that have been used to provide primary and secondary frequency control so the system can be prepared for the next event.

These three stages of control work in a complementary fashion to arrest, stabilize, and restore system frequency in response to energy and load imbalance. Immediately following a loss of generation event, all synchronous rotating machines in the system provide inertial responses. During this stage, all synchronous generators supply extra active power in proportion to their relative size. Primary frequency response then kicks in and seeks to arrest and stabilize frequency before shedding any portion of firm load. This stage is known as the "arresting period" in Fig. 5.1. Once system frequency stops declining, primary frequency control will stabilize it by continuing delivering response. After frequency being stabilized, the secondary frequency control takes action to restore system frequency back to normal operational range. Eventually, tertiary control will restore the reserved capacity that have provided primary frequency control by dispatching other generators and free up the reserve.

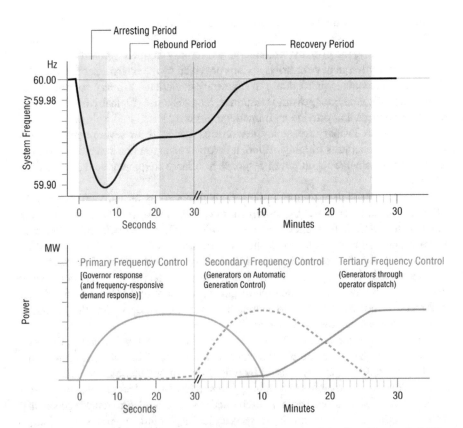

Fig. 5.1 Primary/secondary/tertiary frequency control characters and timeline. (Source: Ref. [4])

2   Impact of Renewable Resource over Inertia and Primary Frequency Control

Future power systems will see significant growth of renewable generation resources (RGSs). In regions such as Ireland and Texas, power generated by RGSs has exceeded 40% of the system load. In many regional grids, fossil fuel units (e.g., coal power plants) are quickly being displaced by inertia-less RGSs (primarily wind and solar), which makes the frequency control a demanding work. In [3–6], the authors demonstrate that the large-scale integration of RGSs leads to decline in system inertia, causing a significant reduction of the primary frequency control (PFC) capability. Electric grids in North America have witnessed this deterioration in the PFC capability over the past decades. Recently, the deterioration is accelerating because the large-scale adoption of wind and solar energies is becoming a reality. Similar studies conducted for the Ireland grid also show that if no grid enhancement is implemented soon, the risk of constraining or curtailing a significant amount of wind generation would be very high because of the lack of adequate PFC [7, 8]. Therefore, whether or not those low inertia interconnected power systems can procure and maintain adequate frequency response reserve (FRR) for providing PFC to respond to credible contingencies is becoming a serious concern [9, 27–31].

Alternative resources for providing FRR include synthetic inertia from wind turbines, energy storage, and demand response [24, 26]. Synthetic inertia from wind turbines cannot sustain the response unless a portion of wind power is spilled [10, 11]. Fast-acting energy storage systems such as batteries or flywheels can quickly stabilize the grid frequency after a disturbance is detected, but their costs are still prohibitively high [12]. Traditionally, demand response is used mainly for peak shaving and load shifting. Recent advancements in communication, sensor, and automation technologies allow fast responsive load to provide fast, high-quality grid services such as load following and regulation [13]. Simulation results and pilot projects have shown that commercial building and residential home loads can effectively respond to under-frequency events such as the loss of large generators [14, 15]. Another advantage for demand response to be frequency responsive is its low cost in the provision of service, which only requires a local device installed to detect frequency deviations and to interrupt the load when needed. Thus, introducing fast-acting load-side resources for FRR is both technically possible and economically efficient.

However, existing studies [16–23] on primary frequency market design, which have built a framework for hosting services provided only by generation resources, did not consider the role of demand response in such a market. The basic principles of co-optimizing energy and primary, secondary, and tertiary reserves are discussed in [21]. In [16, 17], the authors propose a market design mechanism for incentivizing synchronous generators to provide primary frequency response. A simplified dynamic model is introduced to determine the minimum spinning reserve requirement that is used as part of the constraints in economic dispatch for a pool-based power market [18–20, 22]. The problem formulation of accounting for primary service constraints in unit commitment is described in [23]. Those studies focused primarily on the provision of primary frequency response from synchronous generators without considering the participation of other viable resources.

The participation of load resources in the ancillary service market is promising as it can improve both reliability and efficiency. However, there are several challenges associated with the integration of load resources into the primary frequency reserve market. The main difficulty lies in the different performance among a variety of resources in the provision of the primary frequency response. The governor responses from the synchronous machines are continuous and slow, while the actions from load resources in the response to large frequency deviations are discrete and fast. In addition, when introducing the load-provided FRR to the PFC market, it is desirable to know the optimal redistribution of the FRR requirement between loads and generators and to understand how the energy and ancillary service market prices will be affected once loads can bid into both markets. To this end, dramatic changes to existing market mechanism are needed to ensure that a primary frequency reserve market is open to competition, efficient, and nondiscriminatory.

Without the changes required, the existing market design can be a barrier to accommodating load-side resources for providing FRR. First, in some regions, e.g., in the Western Electricity Coordinating Council, the provision of FRR is still mandatory, so there is little monetary incentive for other resources to provide such services [3]. Second, different resources have different response speeds, and the sustainability in maintaining the responses is also different [14, 15]. Thus, the contribution from each resource for stabilizing the frequency excursions needs to be quantified so that different resources with different performance can be harmonized. Third, the most efficient resources should be awarded most so that the market can send an incentive price signal to eventually improve the market efficiency. This is essential when designing a market mechanism for the provision of FRR such that the load and synchronous machines can compete fairly in a single market and be compensated based on their performance.

The rest of this chapter will first introduce the mathematical formulation of a novel primary frequency reserve market where the load resources can participate and then demonstrate the merits and implications of this market through a case study.

3 Co-optimization of Energy and FRR in Day-Ahead Market

3.1 Day-Ahead Market Co-optimization Model

A new day-ahead energy, inertia, and reserve co-optimization formulation is proposed in which the FRR requirement can be met by both primary frequency reserve (PFR) from synchronous generators and fast response reserve (FFR) from load. The proposed work reported is an extension of the previous work which studied the day-ahead market co-optimizing energy and reserves [32–38]. Compared to [31], this formulation can explicitly consider the inter-dependency between the commitment decision, inertia, and PFR/FFR. Unlike the study in [39], the work conducted here determines the award of FFR in the framework of the unit commitment, while it does not require the dynamic simulation as part of the optimization problem, which makes the problem more tractable.

The PFR/FFR requirement can be described by (5.1)–(5.2). The equivalency ratio, m, is the performance metric for FFR and allows PFR to be substituted by FFR without compromising the system primary frequency control capability. Thus, the minimum amount of the FRR requirement becomes

$$m \cdot P_{FFR} + P_{PFR} \geq FRR_{min} \qquad (5.1)$$

$$P_{PFR} \geq PFR_{min} \qquad (5.2)$$

where

$$m = \frac{P_{PFR}}{P_{FFR}}$$

As indicated above, the equivalency ratio, m, explicitly accounts for the different performance between FFR and PFR, and the formulation modeled does not limit the provision of FRR service to PFR only.

The day-ahead market energy and reserve co-optimization problem is a security-constrained unit commitment problem. After considering the constraints in (5.1)–(5.2), the objective function is to maximize the social welfare which is the sum of demand benefits based on demand bids minus energy costs based on generators' three-part offers, reserve costs based on reserve offers from both generating units and demands, and unserved reserve cost based on penalty prices as shown in (5.3). The startup cost is a function of the number of hours the generator has been turned off. The constraint (5.4) implicitly determines the startup costs of generating units in each hour during the optimization process. q represents the index of time segments in stepwise startup curves. STC_{it} will be zero resulting from the optimization process if the unit does not change its status from off to on at hour t.

Formulations (5.5)–(5.8) represent unit commitment status coupling constraints. The startup and shutdown indicators are relaxed in this study to be continuous variables in (5.8) in order to accelerate the computational speed of solving the mixed-integer programming problem by using the branch and cuts strategy [38]. Formulations (5.5)–(5.7) will restrict these continuous variables to be binary value. Formulations (5.9)–(5.11) represent the operational characteristics of individual thermal units, such as ramp rate and minimum on/off time (5.24)–(5.26). Ramp rate limits restrict the difference of power generations in two adjacent hours. Minimum on/off time represents that the generator has to stay online/offline for several hours before it is turned off/on again. Constraints (5.12)–(5.16) define the bounds for the generations and the reserves for each unit. Only the unit that is offline and has quick start capability within 30 minutes can provide non-spinning reserve as shown in (5.17). Constraint (5.18) shows that the frequency response reserve from a load resource must be less than its maximum power consumption that is the highest point in its bid curve. The essential scheduling problem in day-ahead market is to balance cleared energy supply with demand, so hourly generation and cleared demand must satisfy the power balance constraint (5.19). Constraint (5.20) represents transmission constraints under normal or contingency condition. Usually, the network security check is separated from unit commitment problem. Only activated transmission constraints will be added into security-constrained unit commitment constraints. Constraints (5.21)–(5.24) denote the traditional regulation up/down and non-spinning reserve requirement constraints in system-wide. The penalty prices for unserved reserve are usually high so that reserve will be brought in normal condition.

The minimum requirement of PFR is given in formulation (5.25), which implies that at least some synchronous generating units should be committed to contribute their inertias to the system. The overall FRR requirement $Rfrr_t$ is a function of unit commitment and overall inertias of generating units as shown in the right-hand side of (5.26). The equivalent ratio α_t in (5.27) is a measure scale of FFR performance so that FFR can be equivalently replaced by PFR without sacrificing the performance of PFC capability. Equation (5.27) indicates that the FFR requirement can be determined by the overall FRR need, the equivalency ratio, as well as the amount of PFR. FRR requirement-inertia and equivalent ratio-inertia relationship and in (5.26)–(5.27) are a nonlinear function, and formulation (5.27) includes bilinear terms which is a product of two variables so that the model (5.3)–(5.27) is a mixed-integer nonlinear programming which is not effectively solved via commercial solvers, in a reasonable computational time, especially when applied to large-scale power systems. In the next section, we will show how to linearize the FRR requirement-inertia and the ratio-inertia curves and transform bilinear terms by using the linear formulation with big M constraints. In this way, the mixed-integer nonlinear programming model can be reformulated into a mixed-integer linear or quadratic programming model.

$$
\begin{aligned}
\text{Max} \sum_{j\in L}\sum_{t\in T} & \left[Ce_{j,t}\left(L_{j,t}\right) - Cffr_{j,t} \cdot FFR_{j,t} \right] \\
& - \sum_{i\in G}\sum_{t\in T} [STC_{it} + Cf_i \cdot I_{i,t} \cdot LSL_i + Ce_{i,t}(P_{i,t}) \\
& + Crup_{i,t} \cdot RUP_{i,t} + Crdn_{i,t} \cdot RDN_{i,t} \\
& + Cnsr_{i,t} \cdot NSR_{i,t} + Cpfr_{i,t} \cdot PFR_{i,t}] \\
& - \sum_{t\in T} [Nrup_t \cdot RUPN_t + Nrdn_t \cdot RDNN_t \\
& + Nnsr_t \cdot NSRN_t + Npfr_t \cdot PFRN_t \\
& + Nfrr_t \cdot FRRN_t]
\end{aligned}
\tag{5.3}
$$

s.t.

$$\mathrm{STC}_{it} \geq \mathrm{Csu}_{i,\mathrm{qc}} \cdot \left[Y_{i,t} - \sum_{n=1}^{\min(t,q)} I_{i,t-n} \right], \mathrm{STC}_{it} \geq 0 \qquad \forall i \in G, t \in T \qquad (5.4)$$

$$1 - I_{i,t-1} \geq Y_{i,t} \qquad \forall i \in G, t \in T \qquad (5.5)$$

$$I_{i,t-1} \geq Z_{i,t} \qquad \forall i \in G, t \in \mathrm{T} \qquad (5.6)$$

$$I_{i,t} - I_{i,t-1} = Y_{i,t} - Z_{i,t} \qquad \forall i \in G, t \in T \qquad (5.7)$$

$$0 \leq Y_{i,t}, Z_{i,t} \leq 1, I_{i,t} \in \{0,1\} \qquad \forall i \in G, t \in T \qquad (5.8)$$

$$I_{i,t} \geq \sum_{\tau = \max\{1, t - \mathrm{MT}_{\mathrm{on},i}+1\}}^{t} Y_{i,\tau} \qquad \forall i \in G, t \in T \qquad (5.9)$$

$$1 - I_{i,t} \geq \sum_{\tau = \max\{1, t - \mathrm{MT}_{\mathrm{off},i}+1\}}^{t} Z_{i,\tau} \qquad \forall i \in G, t \in T \qquad (5.10)$$

$$- \mathrm{RD}_i \leq P_{i,t} - P_{i,t-1} \leq \mathrm{RU}_i \qquad \forall i \epsilon G, t \in T \qquad (5.11)$$

$$P_{i,t} + \mathrm{RUP}_{i,t} + \mathrm{PFR}_{i,t} \leq \mathrm{HSL}_i \cdot I_{i,t} \qquad \forall i \epsilon G, t \in T \qquad (5.12)$$

$$P_{i,t} - \mathrm{RDN}_{i,t} \geq \mathrm{LSL}_i \cdot I_{i,t} \qquad \forall i \epsilon G, t \in T \qquad (5.13)$$

$$0 \leq \mathrm{RUP}_{i,t} \leq \overline{\mathrm{RUP}}_i \cdot I_{i,t} \qquad \forall i \epsilon G, t \in T \qquad (5.14)$$

$$0 \leq \mathrm{RDN}_{i,t} \leq \overline{\mathrm{RDN}}_i \cdot I_{i,t} \qquad \forall i \epsilon G, t \in T \qquad (5.15)$$

$$0 \leq \mathrm{PFR}_{i,t} \leq \overline{\mathrm{PFR}}_i \cdot I_{i,t} \qquad \forall i \epsilon G, t \in T \qquad (5.16)$$

$$0 \leq \mathrm{NSR}_{i,t} \leq \mathrm{QSC}_i \cdot (1 - I_{i,t}) \qquad \forall i \epsilon G, t \in T \qquad (5.17)$$

$$0 \leq \mathrm{FFR}_{j,t} \leq L_{j,t} \leq \mathrm{MPC}_{j,t} \qquad \forall j \epsilon D, t \in T \qquad (5.18)$$

$$\sum_{i \epsilon G} P_{i,t} = \sum_j L_{d,t} \qquad \forall t \in T \qquad (5.19)$$

$$- \overline{\mathrm{PL}}_l \leq \sum_{i \in G} \mathrm{SF}_{l,i} P_{i,t} + \sum_{j \in L} \mathrm{SF}_{l,j} L_{j,t} \leq \overline{\mathrm{PL}}_l \qquad \forall l \in B, t \in T \qquad (5.20)$$

$$\mathrm{RUPN}_t, \mathrm{RDNN}_t, \mathrm{NSRN}_t, \mathrm{NPFR}_t, \mathrm{FRRN}_t \geq 0 \qquad (5.21)$$

$$\mathrm{RUPN}_t + \sum_{i \epsilon G} \mathrm{RUP}_{i,t} \geq \mathrm{Rrup}_t \qquad \forall t \in T \qquad (5.22)$$

$$\mathrm{RDNN}_t + \sum_{i \epsilon G} \mathrm{RUP}_{i,t} \geq \mathrm{Rrdn}_t \qquad \forall t \in T \qquad (5.23)$$

$$NSRN_t + \sum_{i \epsilon G} NSR_{i,t} \geq Rnsr_t \qquad \forall t \in T \qquad (5.24)$$

$$\text{NPFR}_t + \sum_{i \in G} \text{PFR}_{i,t} \geq \text{Rpfr}_t \qquad \forall t \in T \qquad (5.25)$$

$$\text{FRRN}_t + \text{FRR}_t \geq \text{Rfrr}_t \left(\sum_{i \in G} I_{i,t} \cdot H_i \cdot S_i \right) \qquad \forall t \in T \qquad (5.26)$$

$$\text{FRR}_t = \sum_{i \in G} \text{PFR}_{i,t} + \alpha_t \left(\sum_{i \in G} I_{i,t} \cdot H_i \cdot S_i \right) \cdot \left(\sum_{j \in D} \text{FFR}_{j,t} \right) \qquad \forall t \in T \qquad (5.27)$$

where:

i	Index of generating units
j	Index of loads
t	Index of time periods
s	Index of segments
q	Index of segment in stepwise startup curves
l	Index of transmission branches
P	Cleared energy
L	Cleared demand
I, Y, Z	Binary indicators for unit on/off, startup, and shutdown
δ	Binary variables indicating a segment in a linearized curve is activated
STC	Startup cost of a generating unit
RUP	Regulation up reserve of generating units
RDN	Regulation down reserve of generating units
NSR	Non-spinning reserve of generating units
PFR	Cleared frequency response reserve from primary frequency response of generating units
FFR	Cleared frequency response reserve from fast frequency response of loads
FRR	Total cleared frequency response reserve
RUPN	Not served regulation up reserve
RDNN	Not served regulation down reserve
FRRN	Not served frequency response reserve
NSRN	Not served non-spinning reserve
PFRN	Not served primary frequency response reserve
Inx	Inertia value of a segment in a linearized curve
β	Variables to replace bilinear terms
α	Equivalent ratio between FFR and PFR, depending on the total inertias of committed generation units
Rfrr	Total requirement of frequency response reserve, depending on the total inertias of committed generation units
Ce	Energy cost/benefit curve based on energy offers/bids
G	Set of generating units
D	Set of demands

T	Set of time periods
B	Set of transmission branches
N	Set of segments
Csu	Step constant in the startup cost curve of a generating unit
Cf	Minimum energy price
Nrup, Nrdn	Penalty price for unserved regulation up and down
Nfrr, Nnsr	Penalty price for unserved frequency responsive reserve and non-spinning reserve
Npfr	Penalty price for unserved frequency response reserve and primary frequency response reserve from generating units
Rrup, Rrdn	Total requirement of regulation up/down reserve
Rnsr	Total requirement of frequency responsive reserve and non-spinning reserve
Rpfr	Total requirement primary frequency response reserve from generating units
LSL, HSL	Low and high sustainable limits of generating units
MPC	Maximum power consumption for a demand
MT_{on}, MT_{off}	Minimum on/off time of generating units
RU, RD	Maximal ramp up/down limits per hour of generating units
QSC	Quick start capacity of generating units in 30 minutes
H	Inertial constant
S	Rated power of generating units
SF	Generator/load shift factor of power network
$\overline{PL}$	Capacity limit of a transmission branch
$\overline{RUP}, \overline{RDN}, \overline{PFR}$	Upper bounds of available regulation up/down reserves and PFR of generating units
Ratio	The step value of the ratio-inertia curve
RFRR	The point value in vertical scale of the FRR requirement-inertia curve
In	The inertia value of the linearized curves
M	A larger positive number

3.2   Solution of Day-Ahead Market Co-optimization

Discrete points in the FRR requirement-inertia curve and the equivalent ratio-inertia curve are obtained from dynamic simulation as shown in Table 5.1.

To model it in the market operations, we approximate the equivalent ratio-inertia and the FRR requirement-inertia relationships by using stepwise linear curve and piecewise linear curve as displayed in Figs. 5.2 and 5.3.

In order to represent linearized curves in optimization model, formulations (5.26)–(5.27) are reformulated by (5.28)–(5.32). We introduce additional binary variables to represent the activation of a segment in piecewise linear and stepwise curves. The constraint (5.32) denotes that the current segment must be activated if

Table 5.1 Minimum FFR requirement and equivalency ratios

Case no.	Inertia (GW·s)	FRR$_{min}$(MW)	Equivalency ratio (m)
1	239	5200	2.2
2	271	4700	2.0
3	304	3750	1.5
4	354	3370	1.4
5	403	3100	1.3
6	459	3040	1.25
7	511	2640	1.13
8	556	2640	1.08
9	593	2240	1
10	631	2280	1
11	664	2140	1
12	700	2140	1

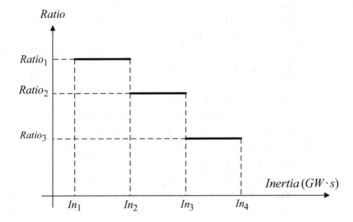

Fig. 5.2 Stepwise equivalent ratio-inertia curve

the next segment is activated. Formulation (5.31) defines the upper bound and the lower bound of inertia value of a segment. If the next segment is activated, the current segment must be binding to its maximum value. If the current segment is not activated, the current segment must be equal to zero. Equation (5.30) represents that the total system inertia is equal to the summation of the inertia value of all segments. The FRR requirement for each delivery hour Rfrr$_t$ is equivalent to the right-hand side of (5.29) where IR$_s$ represents the slope of a segment in a piecewise linear curve. The stepwise ratio curve can be represented by the bilinear terms which are the products of the binary variables corresponding every segment and the continuous variables of cleared FFR as shown in (5.28).

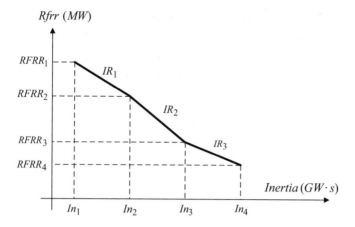

Fig. 5.3 Piecewise linear FRR requirement-inertia curve

$$\text{FRR}_t = \sum_{i \in G} \text{PFR}_{i,t} + \left[\sum_{s \in N} \text{Ratio}_s \cdot (\delta_{s,t} - \delta_{s+1,t}) \right] \left(\sum_{j \in D} \text{FFR}_{j,t} \right) \qquad \forall t \in T$$

$$(5.28)$$

$$\text{FRRN}_t + \text{FRR}_t \geq \text{RFRR}_1 + \sum_{s \in N} \text{IR}_s \cdot \text{Inx}_{s,t} \qquad \forall t \in T \qquad (5.29)$$

$$\text{In}_1 + \sum_{s \in N} \text{Inx}_{s,t} = \sum_{i \in G} I_{i,t} \cdot H_i \cdot S_i \qquad \forall t \in T \qquad (5.30)$$

$$(\text{In}_{s+1} - \text{In}_s) \cdot \delta_{s+1,t} \leq \text{Inx}_{s,t} \leq (\text{In}_{s+1} - \text{In}_s) \cdot \delta_{s,t} \qquad \forall s \in N, t \in T \qquad (5.31)$$

$$\delta_{s,t} \geq \delta_{s+1,t}, \delta_{s,t} \in \{0,1\} \qquad \forall s \in N, t \in T \qquad (5.32)$$

However, due to the bilinear terms, the optimization model (5.3)–(5.25) and (5.28)–(5.32) still cannot be solved by commercial solvers effectively. The way to solve this problem is to expand the feasible region by the big M method. We reformulate the bilinear terms in (5.28) by using some linear big M constraints as described in (5.33)–(5.35). If the binary variables $\delta_{s,t} - \delta_{s+1, t}$ are equal to one, the constraints (5.34)–(5.35) are relaxed. Otherwise, β_t is binding to the linear term, $\text{Ratio}_s \cdot \sum_{j \in D} \text{FFR}_{j,t}$. The choice of the big M value is crucial to the computational speed of branch and cuts process to solve the proposed mixed-integer linear programming model (5.3)–(5.25) and (5.29)–(5.35). Setting big M too small can lead to infeasible or suboptimal solutions. Therefore, the value of big M will typically have to be rather large in order to exceed the largest activity level. When big M is large, the solver may discover that the feasible region of relaxed integer programming problem is also large. It can increase the upper bound of the mixed-integer programming (MIP) (if the objective is to maximize the social welfare) to a significant level, and thereby, the MIP gap is hardly reduced. In order to make the problem tighter, the value of big M should be as small as possible and exceed the largest activity level. Therefore, we chose $M_{s,t} = \text{Ratio}_s \cdot \sum_{j \in D} \text{MPC}_{j,t}$ in this chapter to make the problem tighter. In practice, since only a portion of load resources may provide FFR, we can reduce the big M to a reasonably small value.

$$\text{FRR}_t = \sum_{i \in G} \text{PFR}_{i,t} + \beta_t \qquad \forall t \in T \qquad (5.33)$$

$$\beta_t \geq \text{Ratio}_s \cdot \sum_{j \in D} \text{FFR}_{j,t} - M_{s,t} \cdot (1 - \delta_{s,t} + \delta_{s+1,t}) \qquad \forall s \in N, t \in D \qquad (5.34)$$

$$\beta_t \leq \text{Ratio}_s \cdot \sum_{j \in D} \text{FFR}_{j,t} + M_{s,t} \cdot (1 - \delta_{s,t} + \delta_{s+1,t}) \qquad \forall s \in N, t \in D \qquad (5.35)$$

3.3 Case Studies

The computational study is based on the modified IEEE-118 bus system [23]. The modified IEEE-118 bus power system consists of 54 thermal generating units, 118 buses, and 186 transmission lines. The total installed capacity of 54 thermal generating units is scaled up to 60,000 MW installed thermal generation capacity at ERCOT. The susceptances and thermal rates of transmission branches are also increased in proportion. In this chapter, we focus on the modeling of energy and reserve schedule problem with considering high penetration of renewables. Six identical wind farms are added to buses 11, 15, 54, 59, 80, and 90. It is assumed that all wind power offers are $0.01/MWh. The inertia constants of generating units are given in Fig. 5.4. The maximum capacities for the regulation up reserve, the regulation down reserve, and the PFR reserve of generating units are set as 5%, 5%, and 20% of their high sustainable limit (HSL). The offer prices of the regulation up reserve, the regulation down reserve, the non-spinning reserve, and the PFR reserve are set as 33.3%, 33.3%, 10%, and 20% of the prices of their first energy offer segment. The regulation up/down and non-spinning reserve requirements for different hours in March 2017 can be found in the public market information website of ERCOT [25]. The minimum PFR requirement and FFR requirement are related to the system inertia, which are listed in Table 5.1.

We assume that there are three load resources in the day-ahead market. The peak capacity of the load resources are listed in Table 5.2 at hour 19. The load resources are elastic load with bidding prices $90/MWh, $35/MWh, and $6/MWh. The percentages of the capacities of the load resources at different hours are listed in Table 5.3. We assume that all load resources at different hours have the same bidding prices and FFR offer prices.

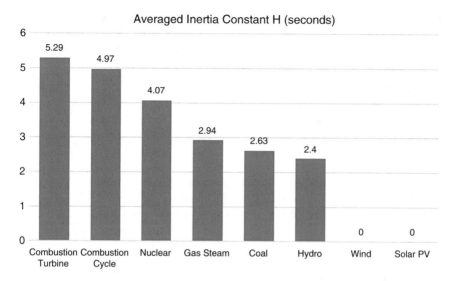

Fig. 5.4 Inertia constant of different generating units

Table 5.2 Bidding parameters and capacities of load resources

Hour 19 (peak)	Load capacity (MW)	Load bid ($/MWh)	FFR offer ($/MW)
L1	29,406	90	45
L2	8822	35	5
L3	2941	6	3

Table 5.3 Capacity percentages of load resources

Hour	Load capacity percentage (%)	Hour	Load capacity percentage (%)	Hour	Load capacity percentage (%)
1	79	9	76	17	86
2	75	10	79	18	99
3	72	11	81	19	100
4	71	12	83	20	98
5	71	13	84	21	96
6	72	14	84	22	92
7	74	15	83	23	86
8	75	16	82	24	80

Three scenarios are simulated, which are described as follows.

Case A—Low penetration of wind generation and low offer prices of FFR. The peak
wind power generation in this base case is 3598 MW at hour 4 so that the wind
power penetration is about 13% at that hour. The hourly maximum wind generation profiles are shown in Table 5.4. The FFR offers of load resources are given
in Table 5.3.

Table 5.4 Total high sustainable limits of low penetration of wind generation

Hour	Wind HSL (MW)	Hour	Wind HSL (MW)	Hour	Wind HSL (MW)	Hour	Wind HSL (MW)
1	3246	7	2230	13	923	19	2239
2	3587	8	2201	14	611	20	2689
3	3530	9	1764	15	1020	21	2903
4	3598	10	1655	16	1056	22	3177
5	3037	11	1243	17	1454	23	3283
6	2631	12	1261	18	1852	24	3319

Table 5.5 Total high sustainable limits of high penetration of wind generation

Hour	Wind HSL (MW)	Hour	Wind HSL (MW)	Hour	Wind HSL (MW)	Hour	Wind HSL (MW)
1	12,985	7	11,016	13	12,870	19	12,534
2	14,347	8	11,526	14	11,868	20	11,184
3	14,121	9	10,584	15	11,370	21	11,610
4	14,391	10	11,370	16	12,036	22	12,709
5	12,148	11	12,078	17	12,792	23	13,132
6	10,524	12	12,228	18	13,548	24	13,275

Case B—High penetration of wind generation and low offer prices of FFR. The peak wind power generation in this case is 14,391 MW at hour 4 so that the total wind power penetration is about 50% at that hour. The hourly maximum wind generation profiles are shown in Table 5.5. The capacities of the load resources and their bid and offer prices are the same as Case A.

Case C—High penetration of wind generation and high offer prices of FFR. The wind power generation in this case is the same as Case B. The capacities of load resources are the same as Case A. We increase the FFR offer price of L2 to $10/MWh.

All algorithms are implemented in AMPL and solved with CPLEX 12.5. The test environment is AIX server with four 4.024-GHz CPU processors and 64 GB of RAM. The MIP gap is set as 0.6%.

The social welfare in Case A is $45,378,657.80 that is lower than that $50,218,472.72 in Case B and $50,177,427.09 in Case C, respectively, because Case B and Case C have more wind energy scheduled in the next day. Furthermore, load resource L2 has the lower FRR offer price in Case B than that in Case C. As a consequence, the social welfare in Case B is slightly higher than Case C. The wall clock times to solve the problem with and without FRR constraints (5.29)–(5.35) are shown in Table 5.6. The computational time of original formulation is longer because of more binary variables and big M constraints. We also propose a warm start process before running the full model of market clearing with FRR constraints. The process is shown in Fig. 5.5. Basically, the market clearing model is executed without FRR constraints. Once the model is solved, the system inertia can be

Table 5.6 Computational time

Wall clock time (s)		Case A	Case B	Case C
With FRR constraints	Original formulation	7.6	84.6	318.9
	Warm start	9.0	7.8	7.5
Without FRR constraints		4.3	3.2	3.2

Fig. 5.5 Proposed warm start process

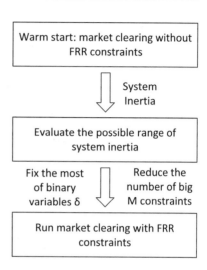

calculated accordingly. Because the total system inertia could not change significantly with FRR constraints, the most of binary variables δ will be fixed except the ones corresponding to the segment of the system inertia resulting from the warm start and its neighboring segments. Therefore, the number of big M constraints will be reduced significantly. With the fixed binary variables and the reduced number of big M constraints, we run the market clearing model with the proposed FRR constraints to obtain the final solution. The computational time with a warm start is much less than that with the original formulation as listed in Table 5.6.

The energy bid price of the load resource L3 is very low, so L3 is not awarded and the other two loads are awarded in their full bidding capacities in all three cases. The total hourly committed capacities for the three cases are shown in Fig. 5.6. Since the wind generation is low in Case A, the total capacity of committed thermal generating units in Case A is the highest among three cases. Case B and Case C have the same cleared wind generation; however, Case C prefers to commit additional units and thus buys PFR from thermal units rather than obtain FFR from LRs. Case C commits more generating capacity to supply the PFR due to the higher FFR offer price of the load resource L2 in Case C.

The total system inertias based on unit commitment for the three cases are given in Fig. 5.7 in which the trend of the three curves is similar with that in Fig. 5.6. It is noted that the higher system inertia can result in a lower FRR requirement and a lower PFR/FFR ratio. As a result, the FRR requirement under high wind penetration condition is higher than that under low wind penetration condition. Therefore, more

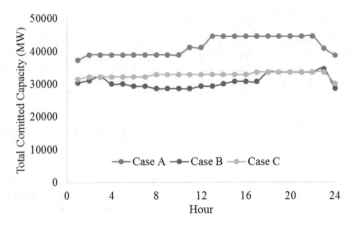

Fig. 5.6 Total committed capacity of thermal generating units

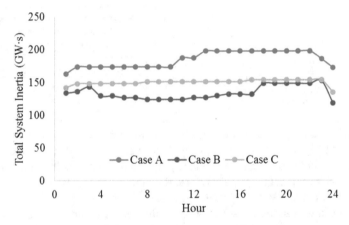

Fig. 5.7 Total system inertia based on unit commitment

FFR are awarded in Case B than Case A, while the cleared PFR are almost the same in the two cases as shown in Figs. 5.8 and 5.9. In Case C, the thermal generating units provide all FRR except the hour 24. The load resource L2 provides additional FRR at hour 24 because all PFR are awarded and committing additional thermal generating units is not economical.

The cleared prices for PFR and FFR are shown in Figs. 5.10 and 5.11. The PFR price is equal to the sum of the dual variables of the constraints (5.25) and (5.26). If the thermal generating units are the marginal resources to supply FRR, the PFR price will reflect the PFR offer of those marginal resources such as hour 1–24 in Case C. If the load resources are marginal resources to supply FRR, both PFR offer and FFR offer will impact the PFR cleared price because there is the minimum PFR requirement constraint (5.25). It is possible that the minimum PFR requirement constraint is binding and the dual variable of (5.25) is not zero, such as the most hours in Case A

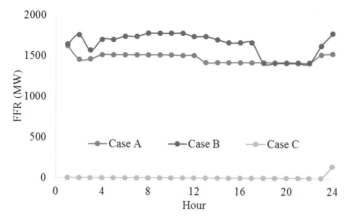

Fig. 5.8 Cleared FFR in day-ahead market

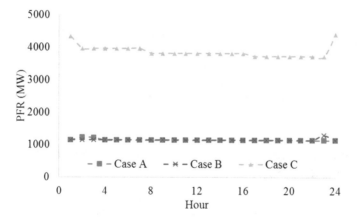

Fig. 5.9 Cleared PFR in day-ahead market

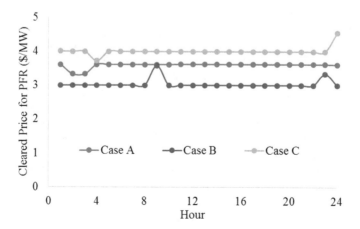

Fig. 5.10 Cleared price for PFR in day-ahead market

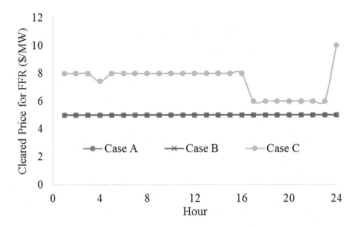

Fig. 5.11 Cleared price for FFR in day-ahead market

and Case B. The cleared FFR price accounts for the marginal offer of FRR and the equivalency ratio between the FFR and PFR. During hour 1 to hour 23 in Case C, there is no awarded FFR since the cleared FFR price is lower than the FFR offer. In Cases A and B, the FFR offer will determine the cleared price of FFR because load resource L2 is the marginal resource to provide FRR.

4   Stochastic Formulations of Co-optimization of Energy and FRR in Day-Ahead Market

Studies of determining reserve requirement other than PFR have been extensively performed in literatures since reserves are of crucial importance to reliable operations of a power system. These approaches can be characterized into two categories: deterministic and probabilistic.

The deterministic approach is rather simple; it is based on a set of pre-determined rules. For example, N-1 rule is intended to protect the system from the single largest generation loss. This is the rule widely used by utilities. The work in [40] also suggested to use 3% of forecast load in addition to 5% of forecast wind to account for wind forecast uncertainties. In [41], the required reserve is dependent upon the permissible degree of demand shedding over a year.

The probabilistic approach incorporates the probabilistic rule and models the rule as a constraint in the unit commitment (UC) problem. The stochastic nature about this approach is twofold.

The first stochastic nature is due to the random forced outages. With the probability of a specific outage being assigned, expected load not served (ELNS) can be explicitly modeled into the optimization problem [42–44]. Bouffard et al. [45, 46] proposed a stochastic security-constrained multi-period market clearing problem with UC. The formulation includes a set of pre-selected generator and line outages with known historical failure rate. The objective function includes the expected social welfare during pre-contingency, the expected reserve deployment cost, and the expected cost of load not served. In such a model, the reserve constraint is no longer a hard constraint, and the violation of reserve is reflected in ELNS.

The second stochastic source is the uncertainty brought by renewables. The growth of wind and solar generation has sparked ideas on the application of stochastic spinning reserve to cover for wind forecast uncertainties in a stochastic UC context [47]. Ruiz et al. [48] argued that a multi-stage stochastic UC model with an explicit reserve constraint is more robust and superior compared with traditional approaches from both economics and reliability perspectives. Papavasiliou et al. [49] developed a two-stage stochastic model for committing and scheduling reserves in the systems with a high penetration of wind. A set of probabilistically weighted wind scenarios are produced for the stochastic UC. The proposed model was tested on a California grid. Botterud et al. [50] divided reserve requirement into two components: regular need and additional need caused by wind generation. The latter is derived from probabilistic forecast and used in a deterministic UC for reserve scheduling. They also feed probabilistic forecast into a stochastic UC to schedule reserve. Ortega-Vazquez and Kirschen [51] proposed an approach to calculate optimal spinning reserve requirement based on a cost/benefit analysis. The optimal amount of spinning reserve should minimize the sum of the expected cost of interruptions and the operating cost with a net load forecast presumingly following a Gaussian distribution. Restrepo and Galiano [52] proposed an approach where reserve should cover the single largest unit plus the demand forecast error within a predefined probability. Lowery et al. [53] applied both deterministic and stochastic UC modeling approaches to examine different deterministic rules. They also investigated the differences between explicit and implicit modeling of reserve constraints.

However, very few attentions have been given to how operating reserves should be properly determined and scheduled so that sufficient PFR can be procured to cope with severe loss-of-generation events. The work in [54] developed a robust UC including frequency stability constraints with a simplified dynamic model representing the frequency response. A similar approach was proposed in [55, 56] to incorporate the analytic frequency deviation into UC while considering a set of scenarios. Using simplified models may reproduce satisfactory frequency responses in a small system, but for a larger, realistic regional grid (e.g., ERCOT), a full network dynamic simulation that considers governor models, contingency location, and load damping effect is needed to calculate the amount of PFR needed to arrest frequency decays.

To summarize, the state-of-the-art approaches suffer a major problem, i.e., they failed to account for uncertainties while accurately considering the system frequency dynamics. To resolve this problem, the work presented here is to develop a novel market model to minimize the cost of jointly clearing the energy, inertia, and PFR in a low inertia power grid. The proposed approach is formulated as a stochastic UC problem while taking into account the forecast uncertainties of renewable generation resources. This method is robust because uncertainties in generation schedules are considered by accounting for the wind generation forecast and their impacts on the amount of PFR needed. In addition, by linearizing the relationship between the inertia and the PFR needs and applying scenario reduction, the proposed approach is computationally efficient and can meet the computation time requirement in market operation.

4.1   Energy, PFR, and Inertia Scheduling Without Uncertainties

The electricity market around the world has evolved to a point where ASs and energy are co-optimized. As such, online generators or controllable load resources providing new RRS will be deployed within seconds for severe frequency drops. The response time of new RRS is well aligned with the delivery timeframe of PFR so that the governor-type PFR response is explicitly procured as a reserve in the market and incentivized by the awards cleared.

Under this new AS framework, a new market model needs to be developed such that the procurement of PFR service and energy can be co-optimized. Moreover, different from any other AS products, PFR varies with respect to the system inertia, which is determined by unit commitment results. As shown in Table 5.1, the amount of PFR increases dramatically when the system inertia decreases. To account for those dependencies, a new scheduling model is developed in this chapter to co-optimize energy, PFR requirements, and the system inertia in both day-ahead and hour-ahead operations.

This section discusses a two-step procedure to include the PFR constraints into the co-optimization problem formulation.

In the *first* step, the methodology introduced in Section 3 is used to derive the PFR requirements using the full ERCOT network model at different net load levels, which is correlated to different system inertia values as shown in Table 5.1. This approach is fundamentally different from pre-determined, reserve rule-based methods because the amount of PFR reserve is calculated in response to realistic system dynamics.

However, the scheduling problem is difficult to solve if PFR requirements under different inertia conditions are directly modeled because the problem is highly nonlinear. Thus, in the *second* step, we linearize the PFR and inertia constraints such that a piecewise linear curve is used to calculate the PFR requirement based on the system inertia. Because the system inertia is a linear function of the generation commitment status represented by binary integer decision variables, the PFR requirement expressed by a piecewise linear function is also linearly dependent upon the binary decision variables. This modification allows the UC problem considering PFR requirements at different system inertia conditions to be formulated as a mixed-integer linear programming (MILP).

Using the piecewise linear function, a series of system inertia conditions and the corresponding PFR requirements can be described as follows. The PFR/inertia function is broken into Ib blocks, $ib \in \{1, \ldots, Ib\}$, with inertia breakpoints $\text{In}(ib)$ and corresponding slopes $\text{sl}(ib)$. Inertia _ min is the minimum inertia level allowed, and PFR _ max is the corresponding PFR requirement. The linearized PFR-inertia relationship is modeled mathematically as the following constraints:

$$\text{Inertia} \sum_{ib} \text{In\_block}(ib, t)_{\min} \tag{5.36}$$

$$\text{PFR} \sum_{ib} \text{sl}(ib) \cdot \text{In\_block}(ib, t)_{\max} \tag{5.37}$$

$$(\text{In}(ib) - \text{Inertiamin}_{\min} \forall t, ib = 1 \tag{5.38}$$

$$(\text{In}(ib) - \text{In}(ib-1)) \cdot bb(ib, t) \leq \text{In\_block}(ib, t) \leq \text{In}(ib) - \text{In}(ib-1) \forall t, 2 \leq ib \leq Ib - 1 \tag{5.39}$$

$$\text{In\_block}(ib, t) \leq (\text{In}(ib) - \text{In}(ib-1)) \cdot bb(ib-1, t) \forall t, ib = Ib \tag{5.40}$$

where:

$bb(ib, t)$ is a binary variable representing the inertia block at which the system inertia falls into. If $bb(ib,t) = 1$, system inertia exceeds $\text{In}(ib)$. Otherwise, system inertia is less than $\text{In}(ib)$.
$\text{In\_block}(ib,t)$ is the inertia for each block ib at time t.
$\text{Inertia\_sys}(t)$ is the system inertia at time t.
$\text{PFR\_syst}(t)$ is the corresponding PFR requirement.

4.2 Energy, PFR, and Inertia Scheduling Under Uncertainties

In an electricity market, actual load consumptions and renewable generation outputs may significantly deviate from their forecasted values. A deterministic UC approach schedules generation reserve capacity using expected values of power demands or generations. However, when there are a large amount of renewable resources in the system, this approach could either under- or over-estimate the uncertainties in net load forecast, resulting in under- or over-provision of PFR.

Stochastic optimization is an advanced methodology to handle uncertainties in the decision-making process. Stochastic UC has been discussed extensively in the past. However, very few literatures have considered the system frequency response under the stochastic optimization framework. Until recently, [57] proposed a modified interval UC model with frequency response constraints. However, it used a simplified dynamic model where the actual system dynamic responses cannot be fully accounted for. In this work, a stochastic UC formulation is proposed so that the

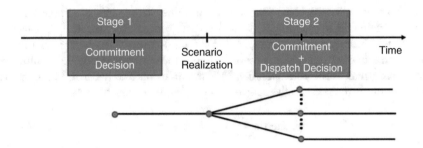

Fig. 5.12 Proposed PFR/inertia scheduling model

actual system frequency behavior can be modeled when accounting for the stochastic nature of renewable outputs.

The proposed stochastic UC can be modeled as a two-stage decision process as shown in Fig. 5.12. Stage 1 represents day-ahead market operation and Stage 2 represents hour-ahead operation. In Stage 1, the commitment decisions only involve the units with status not changeable in the hour-ahead operation (e.g., the units with long startup time or must-run units). In Stage 2, the commitment status of the flexible resources in the 1-hour-ahead operation will be determined.

The here-and-now decision variables in the first stage do not depend on any particular scenario realizations. The wait-and-see decision variables of the second stage depend on the particular scenario realized prior to the operation hour. However, the anticipation of possible outcomes in Stage 2 could have an impact on the decision-making in Stage 1.

The security-constrained economic dispatch (SCED) runs in real time every 5 minutes to produce base point instructions. However, because the outcomes of SCED do not change unit commitment statuses, the SCED process will not change the system inertia and the PFR requirement. Thus, we do not consider the SCED process in this chapter.

The stochastic UC problem with PFR-inertia constraints can be formulated as follows. The objective is to minimize the expected operation cost of energy and PFR:

$$\text{Obj} = \sum_s \text{prob}(s) \cdot \left\{ \sum_{t,i} \begin{bmatrix} C_{\text{energy}}(t,i,s) + C_{\text{pfr}}(i) \cdot \text{pfr}(t,i,s) \\ + \sum_t \text{VOLL} \cdot ll(t,s) \end{bmatrix} \right\} \tag{5.41}$$

Power Balance Constraints:

$$\sum_{i,b} g(t,i,b,s) = d(t,s) - ll(t,s) - w(t,s) + w^{\text{curtail}}(t,s) \qquad \forall t,s \tag{5.42}$$

Generator Capacity Constraints:

$$g(t,i,b,s) \leq g\max(i,b) \cdot x(t,i,s) \qquad \forall t,i,b,s \qquad (5.43)$$

$$\sum_b g(t,i,b,s) \geq g\min(i,b) \cdot x(t,i,s) \qquad \forall t,i,s \qquad (5.44)$$

$$\sum_b g(t,i,b,s) + \mathrm{pfr}(t,i,s) \leq P\max(i) \cdot x(t,i,s) \qquad \forall t,i,s \qquad (5.45)$$

Startup/Shutdown Constraints:

$$y(t,i,s) - z(t,i,s) = x(t,i,s) - x(t-1,i,s) \qquad \forall t,i,s \qquad (5.46)$$

$$y(t,i,s) + z(t,i,s) \leq 1 \qquad \forall t,i,s \qquad (5.47)$$

Ramping Limit Constraints:

$$g(i,t,s) - g(i,t-1,s) \leq \mathrm{ramp}_i^{\mathrm{up}} \qquad \forall t,i,s \qquad (5.48)$$

$$g(i,t-1,s) - g(i,t,s) \leq \mathrm{ramp}_i^{\mathrm{down}} \qquad \forall t,i,s \qquad (5.49)$$

$$\sum_b g(t,i,b,s) = g(i,t,s) \qquad \forall t,i,s \qquad (5.50)$$

Minimum Up-/Downtime Constraints:

$$x(t,i,s) = x_{\mathrm{ini}}(i) \qquad \forall t \leq L_i^{\mathrm{up}} + L_i^{\mathrm{down}}, i, s \qquad (5.51)$$

where

$$L_i^{\mathrm{up}} = \max\{0, \min\{24, (\mathrm{MinUp}(i) - \mathrm{IniUp}(i)) \cdot x_{\mathrm{ini}}(i)\}\}$$

$$L_i^{\mathrm{down}} = \max\{0, \min\{24, (\mathrm{MinDown}(i) - \mathrm{IniDown}(i)) \cdot (1 - x_{\mathrm{ini}}(i))\}\}$$

$$\sum_{tt=t-\mathrm{MinDown}+1}^{t} z(tt,i,s) \leq 1 - x(t,i,s) \qquad \forall t,i,s \qquad (5.52)$$

$$\sum_{tt=t-\mathrm{MinUp}+1}^{t} y(tt,i,s) \leq x(t,i,s) \qquad \forall t,i,s \qquad (5.53)$$

Energy Cost Constraint:

$$\begin{aligned} C_{\mathrm{energy}}(t,i,s) = x(t,i,s) \cdot \mathrm{FC}(i) + y(t,i,s) \cdot \mathrm{Suc}(i) \\ + \sum_b g(t,i,b,s) \cdot VC(i,b) \qquad \forall t,i,s \end{aligned} \qquad (5.54)$$

Inertia/PFR Constraints:

$$\sum_i \mathrm{pfr}(t,i,s)$$

$$\geq \mathrm{PFR\_sys}(t,s) \forall t,s \ (35) \ \mathrm{Inertia} \sum_{ib} \mathrm{In\_block}(ib,t,s)_{\min} \forall t,s \qquad (5.55)$$

$$\mathrm{PFR} \sum_{ib} \mathrm{sl}(ib) \cdot \mathrm{In\_block}(ib,t,s)_{\max} \forall t,s \qquad (5.56)$$

$$\begin{aligned} (\mathrm{In}(ib) - \mathrm{Inertia}_{\min}) \\ \leq \mathrm{In}(ib) - \mathrm{Inertia}_{\min} \end{aligned} \qquad (5.57)$$

$$\begin{aligned} (\mathrm{In}(ib) - \mathrm{In}(ib-1)) \cdot bb(ib,t,s) \leq \mathrm{In\_block}(ib,t,s) \\ \leq \mathrm{In}(ib) - \mathrm{In}(ib-1), \forall t,s, 2 \leq ib \leq Ib-1 \end{aligned} \qquad (5.58)$$

$$\mathrm{In\_block}(ib,t,s) \leq (\mathrm{In}(ib) - \mathrm{In}(ib-1)) \cdot bb(ib-1,t,s) \forall t,s, ib = Ib \qquad (5.59)$$

Since the commitment statuses of inflexible units *{I_nflex}* are Stage 1 variables, they are set to be equal across all scenarios:

$$x(t,i,s) = x(t,i,s+1) \forall t,s \in \{1,..,s-1\}, i \in \{I\_nflex\} \qquad (5.60)$$

$$y(t,i,s) = y(t,i,s+1) \forall t,s \in \{1,..,s-1\}, i \in \{I\_nflex\} \qquad (5.61)$$

$$z(t,i,s) = z(t,i,s+1) \quad \forall t,s \in \{1,..,s-1\}, i \in \{I\_nflex\} \qquad (5.62)$$

where:

b	Index of generator cost curve segments, 1 to B
i	Index of generator, 1 to I
ib	Index of system inertia blocks, 1 to Ib
s	Index of wind forecast scenarios, 1 to S
t	Index of hours, 1 to T
$C_{\mathrm{pfr}}(t,i)$	PFR cost/offering price for generator i at time t (\$/MW)
$d(t)$	Demand at hour t (MW)
FC(i)	No load cost of generator i (\$)
gmax(i,b)	Capacity of segment b of the cost curve of generator i (MW)
gmin(i)	Minimum output of generator i (MW)
In(ib)	Breakpoint ib of the PFR-inertia curve (GW·s)
IniUp(i)	Initial on time of generator i (h)
IniDown(i)	Initial off time of generator i (h)
Inertia _ min	Minimum inertia level allowed (GW·s)
L_i^{up}	Length of time that generator i needs to be on at the beginning of optimization horizon (h)
L_i^{down}	Length of time that generator i needs to be off at the beginning of optimization horizon (h)
MinUp (i)	Minimum uptime of generator i (h)
MinDown(i)	Minimum downtime of generator i (h)

PFR _ max	PFR requirement at selected inertia level (MW)
$P\text{max}(i)$	Maximum output of generator i (MW)
$\text{ramp}_i^{\text{up}}$	Ramp-up limit of generator i (MW/h)
$\text{ramp}_i^{\text{down}}$	Ramp-down limit of generator i (MW/h)
$\text{Suc}(i)$	Startup cost of generator i ($\$$)
$\text{sl}(ib)$	Slope of segment ib of the PFR-inertia curve (MW/GW·s)
$x_{\text{ini}}(i)$	Initial status of generator i (1, on; 0, off)
$x(0,i)$	Initial dispatch point of generator i (MW)
$\text{VC}(i,b)$	Slope of segment b of the cost curve of generator i ($\$$/MW)
VOLL	Value of load loss ($\$$/MWh)
$w(t,s)$	Wind forecast of scenario s at hour t (MW)
$bb(ib,t,s)$	1 (system inertia exceeds In(ib) at hour t in scenario s) or 0 (otherwise)
$C_{\text{energy}}(t, i, s)$	Energy cost of generator i at hour t in scenario s ($\$$)
$g(t,i,b,s)$	Output of generator i on segment b at hour t in scenario s (MW)
$g(t,i,s)$	Output of generator i on segment b at hour t in scenario s (MW)
$\text{In\_block}(ib,t,s)$	Inertia of block ib at hour t in scenario s (GW·s)
$\text{Inertia\_sys}(t,s)$	System inertia at hour t in scenario s (GW·s)
$ll(t,s)$	Load shed at hour t in scenario s (MW)
$pfr(t,i,s)$	PFR responsibility of generator i at hour t in scenario s (MW)
$\text{PFR\_sys}(t,s)$	System PFR requirement at hour t in scenario s (MW)
$x(t,i,s)$	1 (generator i at hour t is on in scenario s) or 0 (otherwise)
$y(t,i,s)$	1 (generator i at hour t is turned on in scenario s) or 0 (otherwise)
$z(t,i,s)$	1 (generator i at hour t is turned off in scenario s) or 0 (otherwise)
$w^{\text{curtail}}(t, s)$	Wind curtailment at hour t in scenario s (MW)

4.2.1   Scenario Generation and Reduction

The proposed model (5.41)–(5.62) requires a set of wind forecast scenarios to properly model wind generation uncertainties. Scenario-based wind forecasting techniques have been discussed extensively in the literature [58–61]. In this study, an optimized autoregressive forecast error generator developed in [62] is used to create the ensemble of wind forecast errors. This wind forecast error generator minimizes the differences between the statistical characteristics of the generated wind forecast error time series and the actual ones. Four matching criteria are considered: the mean, standard deviation, autocorrelation, and cross-correlation between the day-ahead, hour-ahead, and real-time wind forecast error time series. The target statistics are calculated monthly.

The wind forecast error ensembles will be overlaid on top of the wind generation to create a number of scenarios. The computation burden of solving a stochastic UC model increases drastically with the number of scenarios. Therefore, the application of scenario reduction is necessary to keep the size of the problem manageable. A detailed literature review on scenario reduction can be found in [63]. The fast forward selection algorithm [64] is used in this chapter for scenario reduction. The fast forward selection algorithm is an iterative process. It starts with an empty scenario tree. At each iteration, the scenario that minimizes Kantorovich distance between the selected and the initial set is selected. The iteration ends when the target number of scenarios is reached. Then, probabilities of non-selected scenarios are transferred to their closest selected scenarios to form a reduced scenario tree.

4.2.2 Interaction Between Day-Ahead Scheduling and Hour-Ahead Operation

The day-ahead scheduling of energy, PFR, and inertia can be solved using the stochastic form (5.41)–(5.62). During hour-ahead operation, the commitment statuses of a large portion of units (inflexible units) are fixed, and only a subset of generators (flexible units) can alter their online or offline status. Such flexible units are generally either gas or hydro resources.

For stochastic day-ahead scheduling model (5.41)–(5.62), the commitment statuses for inflexible units are Stage 1 variables in the stochastic model so that the following conditions hold:

$$\forall s, \quad x^{DA*}(t, i) = x^{DA*}(t, i, s) \tag{5.63}$$

$$y^{DA*}(t, i) = y^{DA*}(t, i, s) \tag{5.64}$$

$$z^{DA*}(t, i) = z^{DA*}(t, i, s) \tag{5.65}$$

4.2.3 An ERCOT Case Study

The performance of the proposed model is tested on the ERCOT system that comprises a total of 434 thermal units.[1] All units have synthetic cost curves validated against the historical data. Because our main focus is to study the optimal schedule of energy, inertia, and PFR, the network model for addressing the impact of congestion and network losses is not included.

Table 5.7 lists the capacity and inertia information of different generation fleets within ERCOT footprint. Note that gas and hydro units have two sub-categories: flexible and inflexible. Flexible units are subject to commitment status change from day-ahead schedule to hour-ahead schedule, whereas inflexible units' statuses are fixed from day-ahead schedule. The optimization problem is modeled in GAMS and solved by CPLEX. Details on the models used in this study can be found in [65].

---

[1] The current market operation at ERCOT does not co-optimize between energy, inertia, and PFR. The amount of PFR needed for the next day is determined on an annual basis.

Table 5.7 ERCOT generation fleet

		Capacity (MW)	Inertia (GW·s)
Nuclear		5160	23.7
Coal		19,347	61
Gas	Flexible	1968	6.4
	Inflexible	56,349	310.3
Hydro	Flexible	528	1.048
	Inflexible	166	0.7832
Wind		21,412	0

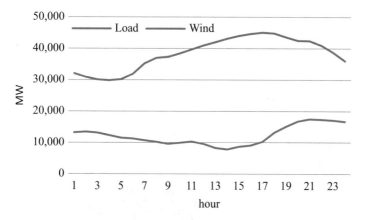

Fig. 5.13 Typical ERCOT daily load and wind profiles in March

4.2.3.1   Deterministic Scheduling Solution

Wind and load profiles are selected from a typical day of March 2017 as shown in Fig. 5.13. For simplicity, network model is ignored. The value of lost load (VOLL) is set to be $9000/MWh. The linearized PFR-inertia relationship for ERCOT system is plotted in Fig. 5.14. The deterministic PFR/inertia constrained UC is solved using the load and wind profile shown in Fig. 5.13. The dispatch result by unit type is shown in Fig. 5.15. Nuclear and coal units are running online to serve base load due to the relatively low cost, while gas and hydro units are committed as intermediate and peaking units. Figure 5.16 shows PFR scheduled and the contribution from different unit types. At each hour, the scheduled PFR and inertia strictly follow the linearized relationship as described in Fig. 5.14. During some peak hours, gas units are allocated with all PFR responsibilities, whereas during other hours, coal and gas units share PFR responsibilities together.

One unique merit of the proposed model is that its decision-making aims to strike a balance between committing resources, meeting PFR requirement, and curtailing wind generation. During high-wind hours, there is a possibility that the proposed model decides to commit additional thermal resources, which would have not been committed otherwise, to increase system inertia, which thus leads to a reduced PFR

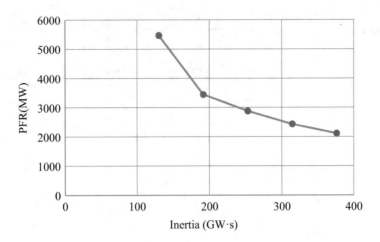

Fig. 5.14 Linearized PFR-inertia relationship of ERCOT system

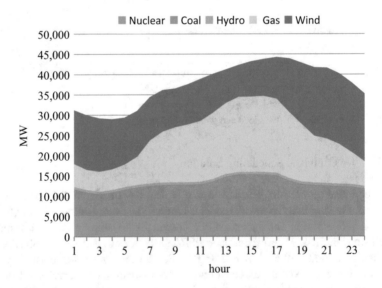

Fig. 5.15 Dispatch by unit type

requirement. The commitment of the additional resources may also result in more wind curtailment. However, as long as the cost reduction from procuring PFR is greater than the cost increase from energy production and instructing additional units online, the proposed model will select such a solution to reduce the total cost.

A sensitivity study of different PFR offering prices is studied to illustrate this unique characteristic. The PFR offering price is assumed to be uniform for all resources, and three levels of PFR offering price ($3/MW, $10/MW, and $18/MW) are investigated. From Fig. 5.17a, b, the model ends up committing

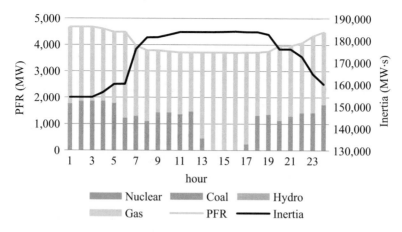

Fig. 5.16 Inertia and PFR schedules

more units as PFR offering price increases. Because of the additional commitments, PFR requirement drops accordingly. As seen in Fig. 5.17c, during the hours when the wind penetration is relatively high (hour 1–5), the over-commitment of the thermal units results in wind curtailment. The highest hourly wind curtailment jumps from 131 MW to 647 MW as PFR offering price increases from $10/MW to $18/MW.

The solution obtained using a single scenario derived in Fig. 5.17 could be very similar to the result obtained from [66, 67]. However, the model described in (5.41)–(5.62) is very computationally tractable because the dynamic simulations are conducted outside of the UC problem. In this regard, the method can easily be extended to account for a set of scenarios as discussed next.

4.2.3.2   Stochastic Scheduling Solution

The method described in Sect. 4.2.1 is used for scenario generation and reduction. First, the statistics of ERCOT historical wind forecast errors for each month are calculated and then normalized with respect to the wind installed capacity. One example of monthly wind forecast error statistics is provided in Table 5.8.

Second, a large number of wind forecast error ensembles are created based on the statistics summarized in Table 5.3 to produce the wind forecast uncertainties. Figure 5.18 shows 100 day-ahead wind forecast scenarios for consecutive 7 days for the ERCOT system (in green) in which each scenario is a combination of one wind forecast error ensemble and the actual wind generation. ERCOT is currently using a third-party numerical weather model to produce the deterministic day-ahead wind forecast each hour for the next 168 hours.

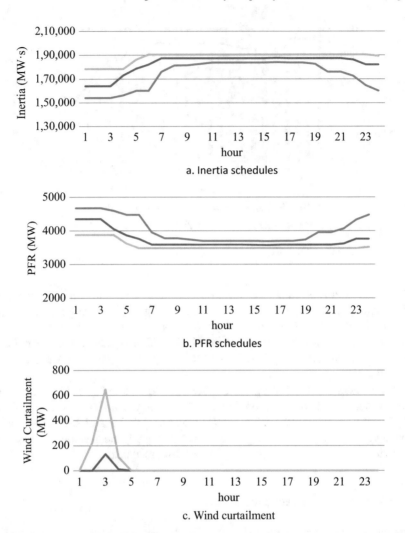

Fig. 5.17 Inertia/PFR scheduling with different levels of PFR offering price (blue, $3/MW; red, $10/MW; green, $18/MW). (**a**) Inertia schedules. (**b**) PFR schedules. (**c**) Wind curtailment

Table 5.8 Wind forecast error statistics

	Standard deviation	Mean	Autocorrelation (t, t-1)	Cross-correlation
Hour-ahead	0.0424	0.0067	0.7514	0.9275
Day-ahead	0.0588	−0.0184	0.8005	0.6654

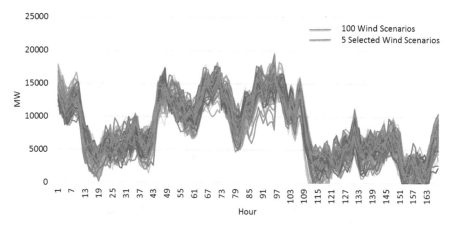

Fig. 5.18 Five reduced day-ahead wind forecast scenarios of a week in March

Third, the fast forward selection algorithm is then applied to those 100 scenarios, and 5 representative scenarios are selected to cover the expected wind variations (in orange) as shown in Fig. 5.18.

The results of stochastic scheduling with PFR/inertia constraint can be derived by solving (5.41)–(5.62) and using the five representative scenarios. Actual system inertia is composed of inertia scheduled in day ahead from inflexible units and the inertia provided in hour ahead from flexible units. As seen in Fig. 5.19, the deterministic approach would schedule less inertia for peak hours in day ahead due to over-forecast of wind generation. In hour ahead, all flexible units are brought online to make up for the energy shortfall. On the other hand, the stochastic model could commit more inflexible units in the day ahead. As a result, not only energy is in shortage but also hour-ahead PFR needs are higher (Fig. 5.19). Committing all flexible resources by the deterministic model is still not able to provide enough capacity for the combined needs of energy and PFR, which eventually results in load shedding. In such a case, stochastic scheduling is more efficient and reliable than deterministic scheduling.

4.2.3.3 Comparison Between Deterministic and Stochastic Scheduling Solutions

One challenge which the grid operator is facing in day-ahead operation is how to effectively schedule resources with the information only available at that time. The past works have demonstrated that stochastic scheduling yields a more cost-efficient result than deterministic scheduling [51]. However, the conclusion could be highly dependent on how the single deterministic scenario is being produced. For example, [51] argues that the proposed stochastic model is more cost-efficient compared to the deterministic model based on a single scenario with a quantile of 0.96. Therefore, it is critical to investigate the outcomes of various scenarios in deterministic

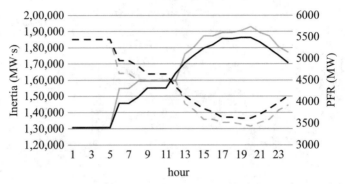

a. Inertia, PFR schedule (red solid line: Inertia-stochastic, red dot line: PFR-stochastic, black solid line: Inertia-deterministic, black dot line: PFR-deterministic)

b. load curtailment (red: stochastic, black: deterministic)

Fig. 5.19 Inertia, PFR schedule, and load curtailment comparison between deterministic and stochastic scheduling. (**a**) Inertia, PFR schedule (red solid line, inertia-stochastic; red dot line, PFR-stochastic; black solid line: inertia-deterministic; black dot line, PFR-deterministic). (**b**) Load curtailment (red, stochastic; black, deterministic)

scheduling so that the comparison of the performance between deterministic and stochastic scheduling can be statistically valid.

To evaluate the performance of the proposed day-ahead stochastic model (5.41)–(5.62), a large number of single scenarios are selected to produce deterministic scheduling results. Thus, a single scenario is randomly drawn from the stochastic scenario pool based on its corresponding probability, which is determined from scenario reduction. For instance, if the probability associated with a wind forecast is 0.6, then there is a 60% of chance that this wind forecast will be chosen. The selection process will be repeated for each day until the end of the study period (7 days). Once each scenario is selected, a deterministic UC model will be executed to calculate the corresponding production cost. The selection process only stops until the fifth percentile and 9fifth percentile of the production cost converge within a 0.01% gap. Figure 5.20 shows both fifth percentile and 9fifth percentile of the production cost for all scenarios as the number of scenarios selected increases.

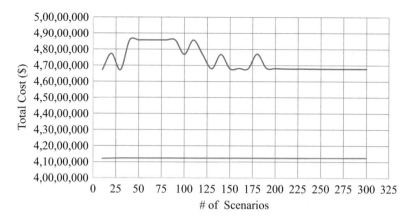

Fig. 5.20 Production cost statistics vs. number of scenarios (blue, fifth percentile; red, 95th percentile)

Table 5.9 Summary of cost statistics

Installed wind capacity (GW)	Stochastic (Million $)	5th percentile deterministic (Million $)	95th percentile deterministic (Million $)	Average deterministic (Million $)
30	38.16	38.13	48.76	40.35
26	41.32	41.24	46.78	42.59
21	45.83	45.72	46.98	46.04

After more than 200 scenarios being selected, almost no difference can be found in the fifth percentile and 95th percentile of the production cost. In this case, 200 scenarios are statistically sufficient to represent the performance of deterministic scheduling.

A cost comparison between the proposed stochastic scheduling model (5.41)–(5.62) and the deterministic scheduling model is then conducted. Table 5.9 summarizes various cost statistics for both approaches.

Figure 5.21 shows a cumulative distribution function (CDF) of the production cost calculated by the deterministic scheduling model in comparison to the stochastic scheduling. The dot vertical line represents the cost calculated from the stochastic scheduling model. The intersection between the CDF function and the dot line indicates how well the stochastic model performs against the deterministic model. In the first case where the wind installed capacity is 30 GW, the stochastic model outperforms deterministic model in terms of the production cost over 90% of the time. Moreover, the production cost of the stochastic model is only slightly higher than the lowest cost calculated from the deterministic model.

The performance comparison between the two scheduling models is also conducted under three levels of wind installed capacities. As shown in Fig. 5.21, the benefit of stochastic scheduling increases along with the installed wind capacity. For 21 GW of wind installed capacity, the stochastic model is more cost-effective

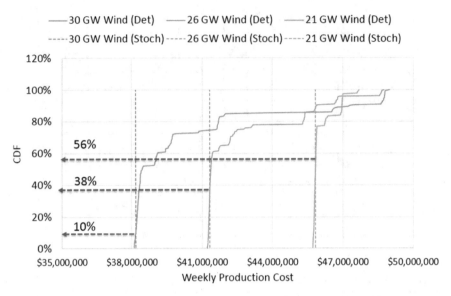

Fig. 5.21 Comparison between deterministic scheduling and stochastic scheduling (solid line, deterministic; dot vertical line, stochastic)

than the deterministic model for 44% of the time. With 26 GW of wind generation integrated to the ERCOT system, the stochastic model incurs a lower cost than deterministic model for 62% of the time. This is attributed to the fact that as a larger amount of wind generation resources can introduce more uncertainties in day-ahead scheduling, the stochastic model can more robustly and efficiently schedule the units.

Another aspect which needs consideration is the computation time. The main factors influencing the computation complexity are the size of the system and the number of scenarios under consideration. The proposed stochastic model yields a runtime under 15 minutes for ERCOT system using five wind forecast scenarios, while the deterministic model takes less than 1 minute to solve. The computation speed of stochastic scheduling model is also tested with more wind forecast scenarios. When the number of wind scenarios is increased to 15, the computation time reaches around 30 minutes. Simulation is carried out in GAMS 24.7.1 using CPLEX 12.6.3 MIP solver with 0.1% duality gap on an Intel Core i7 3.4 GHz processor with 16 GB RAM. If executed on a high-performance computer, the computation time will be much shortened. In contrast, if the methods in [66, 67] were used, a single run of the dynamic simulation for 20 seconds would take minutes to complete. If such dynamic simulation processes are embedded in the optimization problem, the computational time will be significantly longer and may no longer be feasible to be used in the day-ahead scheduling process. From Fig. 5.22, it is demonstrated that the proposed stochastic method is computationally efficient and it can meet the time requirement for both day-ahead scheduling and hour-ahead operations.

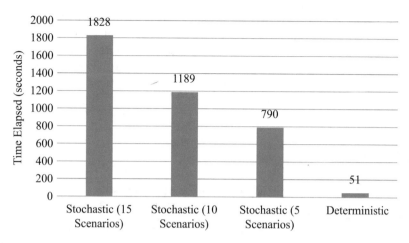

Fig. 5.22 Computation speed comparison

5 Conclusions

Increasing growth of renewable resources could lead to a declined system inertia, which poses a threat to the grid security and reliability. Among various solutions proposed, to maintain sufficient PFC capability is crucial under high penetration of renewable energy condition. However, to allocate such a PFC capacity from the generation side will come at a high cost especially when the online reserve is scarce. Alternatively, load resources can provide a comparable, or even better, primary frequency response. This will benefit a future power grid both reliably and economically by lifting the burden partially from the generation side. The operation paradigm for the future grid will be also shifted if a FRR market is established. This will incentivize these resources which can contribute to the system reliability when needed and reward them based on their performance. In the long term, it will attract more participants to compete, and the market efficiency will be eventually improved.

Based on a case study using ERCOT data and model, the advantages of creating a FRR market are found to be tremendous. Numerical simulations show that PFR and FFR can be substituted with an equivalency ratio, which is largely impacted by system inertia. Since this dependency is nonlinear, it can be approximated as stepwise ratio-inertia curve and piecewise linear FRR requirement-inertia curve. After this simplification, the co-optimization model becomes more tractable. These case studies show the effectiveness of the proposed solution and the correctness of the cleared reserve and prices.

These discussions clearly lay out a path of operating a competitive PFC market to integrate large-scale renewable resources. While this work is primarily focused on frequency responsive LRs, the fundamental principle can be extended to other resources which are also able to respond to the large frequency deviations, such as residential load, energy storages, and synthetic inertia from wind turbines. Active research work is also being pursued to exploit the potential of aggregating a large number of residential load to provide primary frequency control, and the results are very encouraging. As more efficient and less costly resources can respond to the large frequency deviations, the declining inertia driven by the high penetration of renewable resources will not become a concern so that more renewables can be fully integrated to the grid.

References

1. P. Kundur, Power System Stability and Control, McGraw-Hill, 1994.
2. Bevrani, Hassan. Robust power system frequency control. New York: Springer, 2009.
3. E. Ela, M. Milligan, B. Kirby, A. Tuohy and D. Brooks, "Alternative approaches for incentivizing the frequency responsive reserve ancillary service," NREL, March 2012.
4. Eto, Joseph H. "Use of frequency response metrics to assess the planning and operating requirements for reliable integration of variable renewable generation." Lawrence Berkeley National Laboratory (2011).
5. NERC Frequency Response Initiative, April, 2010.
6. Sandip Sharma, Shun-Hsien Huang and NDR Sarma, "System inertial frequency response estimation and impact of renewable resources in ERCOT interconnection," 2011 IEEE Power and Energy Society General Meeting, pp. 1 - 6, San Diego, CA, 24-29 July, 2011.
7. Ronan Doherty, Alan Mullane, Gillian (Lalor) Nolan, Daniel J. Burke, Alexander Bryson, and Mark O'Malley, "An assessment of the impact of wind generation on system frequency control," IEEE Transactions on Power Systems, vol. 25, no. 1, pp. 452–460, Feb. 2010.
8. O'Sullivan, Jon, et al. "Studying the maximum instantaneous non-synchronous generation in an Island system—Frequency stability challenges in Ireland." IEEE Transactions on Power Systems, vol.29, no.6, 2014, pp.2943–2951.
9. Shun-Hsien Huang, John Dumas, Carlos González-Pérez, and Wei-Jen Lee, "Grid security through load reduction in the ERCOT market," IEEE Transactions On Industry Applications, vol. 45, no. 2, pp. 555–559, MARCH/APRIL 2009.
10. Johan Morren, Jan Pierik, and Sjoerd WH De Haan. "Inertial response of variable speed wind turbines," Electric power systems research, vol.76, no.11, 2006, pp. 980–987.
11. Johan Morren et al. "Wind turbines emulating inertia and supporting primary frequency control," IEEE Transactions on Power Systems, vol.21, no.1, 2006, pp.433–434.
12. Uthier Delille, Bruno Francois, and Gilles Malarange. "Dynamic frequency control support by energy storage to reduce the impact of wind and solar generation on isolated power system's inertia," IEEE Transactions on Sustainable Energy, vol.3, no4, 2012, pp.931–939.
13. Ookie Ma, Nasr Alkadi, Peter Cappers, Paul Denholm, Junqiao Dudley, Sasank Goli, Marissa Hummon, Sila Kiliccote, Jason MacDonald, Nance Matson, Daniel Olsen, Cody Rose, Michael D. Sohn, Michael Starke, Brendan Kirby, and Mark O'Malley, "Demand response for ancillary services," IEEE Transactions on Smart Grid, vol. 4, no. 4, pp. 1988 – 1995, Dec. 2013.
14. Angel Molina-García, François Bouffard, Daniel S. Kirschen, "Decentralized demand-side contribution to primary frequency control," IEEE Transactions on Power Systems, vol. 26, no. 1, pp. 411–419, Feb. 2011.
15. Autonomous demand response for primary frequency regulation, Pacific Northwest National Laboratory, 2012.
16. Erik Ela, Vahan Gevorgian, Aidan Tuohy, Brendan Kirby, Michael Milligan, and Mark O'Malley, "Market designs for the primary frequency response ancillary service—Part I: motivation and design," IEEE Transactions on Power Systems, vol. 29, no. 1, pp. 421 – 431, Jan. 2014.

17. Erik Ela, Vahan Gevorgian, Aidan Tuohy, Brendan Kirby, Michael Milligan, and Mark O'Malley, "Market designs for the primary frequency response ancillary service—part II: case studies," IEEE Transactions on Power Systems, vol. 29 , no. 1, pp. 432 – 440, Jan. 2014.
18. J W O'Sullivan and M J O'Malley, "Economic dispatch of a small utility with a frequency based reserve policy," IEEE Transactions on Power Systems, vol. 11, no. 3, pp. 1648–1653, August 1996.
19. J.W. O'Sullivan, M.J. O'Malley, "A new methodology for the provision of reserve in an isolated power system," IEEE Transactions on Power Systems, vol. 14, no. 2, pp. 519–524, May 1999.
20. K. A. Papadogiannis and N. D. Hatziargyriou, "Optimal allocation of primary reserve services in energy markets," IEEE Transactions on Power Systems, vol. 19, no. 1, pp. 652–659, Feb. 2004.
21. F.D. Galiana, F. Bouffard, J.M. Arroyo, J.F. Restrepo, "Scheduling and pricing of coupled energy and primary, secondary, and tertiary reserves," Proceedings of the IEEE, vol. 93, no. 11, pp. 1970–1983, Nov. 2005.
22. R. Doherty, G. Lalor, M. O'Malley, "Frequency control in competitive electricity market dispatch," IEEE Transactions on Power Systems, vol. 20, no. 3, pp. 1588 – 1596, Aug. 2005.
23. J.F. Restrepo, F.D. Galiana, "Unit commitment with primary frequency regulation constraints," IEEE Transactions on Power Systems, vol. 20, no. 4, pp. 1836 - 1842, Nov. 2005.
24. NERC Reliability Standard BAL-003 Frequency Response and Frequency Bias Setting, NERC.
25. ERCOT FAST Consolidated Working Document 5-16-14
26. Third-Party Provision of Primary Frequency Response Service, FERC, November 20, 2015.
27. Pengwei Du, Y.V. Makarov, M. A. Pai, B. McManus, "Calculating individual resources variability and uncertainty factors based on their contributions to the overall system balancing needs," *IEEE Transactions on Sustainable Energy*, vol. 5, no. 1, pp. 323 – 331, Jan. 2014.
28. Pengwei Du, Julia Matevosyan, "Forecast system inertia condition and its impact to integrate more renewables," *IEEE Transactions on Smart Grid*, vol.9, no.2, p. 1531–1533, 2017.
29. P. Du, H. Hui, and N. Lu, "Procurement of regulation services for a grid with high-penetration wind generation resources: a case study of ERCOT," *IET Generation, Transmission & Distribution*, vol.16, no. 10, pp. 4085–4093, 2016.
30. P. Du, Y. Makarov, "Using disturbance data to monitor primary frequency response for power system interconnections," *IEEE Transactions on Power Systems*, vol. 29, no. 3, pp. 1431 – 1432, May 2014.
31. W. Li, P. Du, N. Lu, "Design of a new primary frequency control market for hosting frequency response reserve offers from both generators and loads," *IEEE Transactions on Smart Grid*, vol. 9, no.5, p. 4883–4892, 2017.
32. C. Liu, A. Botterud, Z. Zhou, and P. Du, "Fuzzy energy and reserve co-optimization with high penetration of renewable energy," *IEEE Transactions on Sustainable Energy,* vol. 8, no 2, pp. 782–191, 2017.
33. Wood, A.J., and B.F. Wollenberg, *Power Generation, Operation and Control*, 2nd Edition, John Wiley & Sons, Inc., New York, NY, 1996.
34. Shahidehpour, M., H. Yamin, and Z. Li, *Market Operations in Electric Power Systems,* John Wiley & Sons, Inc., New York, NY, 2002.
35. C. Liu, M. Shahidehpour, Z. Li, and M. Fotuhi-Firuzabad "Component & mode models for short-term scheduling of combined-cycle units," *IEEE Trans. Power Syst.*, vol. 24, pp. 976–990, May 2009.
36. C. Lee, C. Liu, S. Mehrotra and M. Shahidehpour, "Modeling transmission line constraints in two-stage robust unit commitment problem," *IEEE Transactions on Power Systems,* vol. 29, no. 3, pp. 1221 - 1231, 2014.
37. Y. T. Tan and D. L. S. Kirschen, "Co-optimization of energy and reserve in electricity markets with demand-side participation in reserve services." Power Systems Conference and Exposition, 2006. PSCE'06. 2006 IEEE PES. IEEE, 2006.

38. D. Rajan, and S. Takriti, "Minimum up/down polytopes of the unit commitment problem with start-up costs, "*IBM Research Report*, June 2005.
39. Bhana, Rajesh, and Thomas J. Overbye, "The commitment of interruptible load to ensure adequate system primary frequency response," IEEE Transactions on Power Systems, vol.31, no. 3, 2016, pp. 2055–2063.
40. The Western Wind and Solar Integration Study, National Renewable Energy Laboratory, 2010 May.
41. Doherty R, O'malley M, "A new approach to quantify reserve demand in systems with significant installed wind capacity," IEEE Transactions on Power Systems, vol. 20, no.2, pp.587–95, 2005 May.
42. Anstine LT, Burke RE, Casey JE, Holgate R, John RS, Stewart HG, "Application of probability methods to the determination of spinning reserve requirements for the Pennsylvania-New Jersey-Maryland interconnection," IEEE Transactions on Power Apparatus and Systems, vol.82, no.68, p.726–35, 1963.
43. Dillon TS, Edwin KW, Kochs HD, Taud RJ. "Integer programming approach to the problem of optimal unit commitment with probabilistic reserve determination," IEEE Transactions on Power Apparatus and Systems, vol.6, p.2154–66, 1978.
44. Gooi HB, Mendes DP, Bell KR, Kirschen DS, "Optimal scheduling of spinning reserve," IEEE Transactions on Power Systems, vol.14, no.4, p.1485–92, 1999 Nov.
45. Bouffard F, Galiana FD, "An electricity market with a probabilistic spinning reserve criterion," IEEE Transactions on Power Systems, vol. 19, no.1, p:300–307, 2004.
46. Bouffard F, Galiana FD, Conejo AJ. "Market-clearing with stochastic security-part I: formulation," IEEE Transactions on Power Systems, nol.20, no.4, p.1818–26, 2005 Nov.
47. Tuohy A, Meibom P, Denny E, O'Malley M. "Unit commitment for systems with significant wind penetration," IEEE Transactions on power systems, vol.24, no.2, p.592–601, 2009 May.
48. Ruiz PA, Philbrick CR, Zak E, Cheung KW, Sauer PW, "Uncertainty management in the unit commitment problem," IEEE Transactions on Power Systems, vol.24, no.2, p.642–51, 2009 May.
49. Papavasiliou A, Oren SS, O'Neill RP. "Reserve requirements for wind power integration: A scenario-based stochastic programming framework," IEEE Transactions on Power Systems, vol. 26, no.4, p.2197–206, 2011 Nov.
50. Botterud A, Zhou Z, Wang J, Valenzuela J, Sumaili J, Bessa RJ, Keko H, Miranda V. "Unit commitment and operating reserves with probabilistic wind power forecasts," 2011 IEEE Trondheim PowerTech, 2011 Jun 19, pp. 1–7.
51. Ortega-Vazquez MA, Kirschen DS. "Estimating the spinning reserve requirements in systems with significant wind power generation penetration," IEEE Transactions on power systems, vol. 24, no.1, p.114–124, 2008 Nov.
52. Restrepo JF, Galiana FD, "Assessing the yearly impact of wind power through a new hybrid deterministic/stochastic unit commitment," IEEE Transactions on Power Systems, vol. 26, no.1, p. 401–10, 2010 May.
53. Lowery C, OMalley M. "Reserves in stochastic unit commitment: An Irish system case study," IEEE Transactions on Sustainable Energy, vol. 6, no. 3, p.1029–38, 2014 Nov.
54. Prakash V, Sharma KC, Bhakar R, Tiwari HP, Li F, "Frequency response constrained modified interval scheduling under wind uncertainty," IEEE Transactions on Sustainable Energy, vol. 9, no. 1, p.302–10, 2017 Jul.
55. Pérez-Illanes, Felipe, Eduardo Álvarez-Miranda, Claudia Rahmann, and Camilo Campos-Valdés, "Robust unit commitment including frequency stability constraints," Energies, no. 11, p. 957, 2016.
56. Paturet M, Markovic U, Delikaraoglou S, Vrettos E, Aristidou P, Hug G, "Stochastic unit commitment in low-inertia grids," IEEE Transactions on Power Systems, vol. 35, no. 5, p. 3448–3458, 2020.
57. Liu C, Du P, "Participation of load resources in day-ahead market to provide primary-frequency response reserve," IEEE Transactions on Power Systems, vol. 33, no.5, p.5041–5051, 2018.

58. Cassola F, Burlando M. "Wind speed and wind energy forecast through Kalman filtering of numerical weather prediction model output," Applied Energy, vol. 99, p.154–66, 2012.
59. González-Aparicio I, Zucker A. "Impact of wind power uncertainty forecasting on the market integration of wind energy in Spain," Applied Energy, vol. 159, p.334–349, 2015.
60. Pinson P, Nielsen HA, Madsen H, Kariniotakis G. "Skill forecasting from ensemble predictions of wind power," Applied Energy, vol. 86, p. 1326–1334, 2009.
61. Pinson P, Girard R. "Evaluating the quality of scenarios of short-term wind power generation," Applied Energy, vol. 96, p.12–20, 2012 Aug.
62. de Mello PE, Lu N, Makarov Y. "An optimized autoregressive forecast error generator for wind and load uncertainty study," Wind Energy, vol. 14, no.8, p.967–76, 2011 Nov.
63. Park, SangWoo, Qingyu Xu, and Benjamin F. Hobbs. "Comparing scenario reduction methods for stochastic transmission planning," IET Generation, Transmission & Distribution, vol. 13, no. 7, p. 1005–1013, 2019.
64. Antonio J. Conejo, Miguel Carrion, Juan M. Morales, Decision Making Under Uncertainty in Electricity Markets, New York: Springer, 2010
65. Weifeng Li, Primary Frequency Response Ancillary Service in Low Inertia Power Systems, Ph. D. thesis, 2018.
66. Ela E, Gevorgian V, Tuohy A, Kirby B, Milligan M, O'Malley M. "Market designs for the primary frequency response ancillary service—Part I: Motivation and design," IEEE Transactions on Power Systems, vol. 29, no. 1, p.421–31, 2013.
67. Ela E, Gevorgian V, Tuohy A, Kirby B, Milligan M, O'Malley M, "Market designs for the primary frequency response ancillary service—Part II: Case studies," IEEE Transactions on Power Systems, vol. 29, no.1, p. 432–40, 2013 Jun.

Chapter 6
New Ancillary Service Market for ERCOT: Fast Frequency Response (FFR)

1 Introduction

Following Federal Energy Regulatory Commission (FERC) order 888, the US power industry began its restructuring process [1]. Essentially, the competition was introduced among the generation companies after the vertically integrated utility was unbundled [2]. To date, wholesale electricity markets in the United States, with a goal to increase the power system's economic efficiency without compromising its reliability, have been considered successful [3]. In the United States, the market operations are administratively managed by independent system operators (ISOs). While the market design differs across the regions, all of those markets procure and manage an array of ancillary service (AS) products to ensure that they can balance the supply and demand for energy in real time. In this regard, ASs are necessary to support the transmission of energy from resources to loads while maintaining reliable operation of the transmission service provider's transmission system [4]. It is also a prevailing practice to co-optimize the provision of energy and ASs to achieve the most efficient capacity allocation of resources.

AS market has evolved with changes in regulatory policy, technological innovations, and economic conditions since the start of electric restructuring [4–6]. The AS products are designed according to the needs of reliable operation for power systems. Because the different systems may have different reliability needs, AS products could be different among ISOs.

Today, variable energy resources (VERs) have grown dramatically in their installed capacity. While VERs have tremendous benefits, a high penetration of VERs creates a new challenge to maintaining the reliability and security of a power grid [7, 8]. Therefore, moving toward a future power system with a lower carbon footprint, it is necessary to make changes to the AS market design as the new reliability need arises. Recently, some changes in the AS market design have been implemented. For example, California ISO introduced a 5-minute ramping product, and Midcontinent ISO added a 10-minute ramping product to their AS market

© The Author(s), under exclusive license to Springer Nature Switzerland AG 2023
P. Du, *Renewable Energy Integration for Bulk Power Systems*, Power Electronics and Power Systems, https://doi.org/10.1007/978-3-031-28639-1_6

[7]. Both of them are mainly used to mitigate the variability and uncertainties of VERs.

VERs are also connected to the grid through an inverter so that they do not contribute to the system inertia. Power system inertia is defined as the ability of a power system to oppose changes in system frequency due to resistance provided by rotating masses [9]. Inertia is dependent on the amount of kinetic energy stored in rotating masses of synchronously interconnected machines. As VERs do not provide inertia, the large-scale integration of VERs could lead to a lower system inertia. A declining system inertia could result in a difficulty in regulating the system frequency [10–13], which is considered as one of the major barriers to the reliable integration of VERs at the large scale.

Various works have been performed to address the need of improving the secure and reliable operation of a future power grid under the low system inertia conditions. In [14], the frequency stability challenges at ultra-high wind penetrations were examined, and a system non-synchronous penetration ratio was defined to help to identify system operational limits. The authors in [15] proposed a framework for assessing renewable integration limits concerning power system frequency performance using a time-series scenario-based approach. The study in [16] established the instantaneous penetration level limits of VERs. In [17, 18], the frequency response change trend of each US interconnection due to the increasing penetration level of VERs was examined. All of these studies pointed to a great need of the primary frequency response (PFR) capability as the system inertia decreases. PFR denotes the autonomous reaction of system resources to change in frequency [19]. In the past, the main contributor to PFR has been the governor response of synchronous generation units.

Given the significance of PFR, researchers also studied how to incentivize synchronous generators to provide such a PFR capability. The basic principle of scheduling and pricing of coupled energy and primary, secondary, and tertiary reserves is discussed in [20]. A simplified dynamic model is introduced to determine the minimum spinning reserve requirement that is used as part of the constraints in economic dispatch for a pool-based power market [21–23]. The dependency between the system inertia and PFR is approximately taken into account in the scheduling process [24]. More detailed models of the governor responses are adopted in [25, 26] to calculate the pricing of PFR influenced by different dynamic characteristics of the governors. The problem formulation accounting for PFR constraints in unit commitment is described in [27]. Those theoretic studies focused primarily on the provision of PFR from synchronous generators without considering other viable resources. In [28, 29], an energy, inertia, and reserve co-optimization formulation was proposed in which the PFR requirement can be met by synchronous generators and load resources.

However, these theoretic works reported in [20–29] have been performed assuming that no changes are applied to existing AS products. The practical effort to reinvent the AS products is still lacking, but it is imperative to incentivize the resources like batteries to provide fast frequency response capabilities. One exception is the national electricity market (NEM) in Australia, where the 6-second frequency control ancillary service (FCAS) product is used to arrest the frequency decline and the 60-second product is used to stabilize the system frequency [30].

The Electric Reliability Council of Texas (ERCOT) is an ISO serving over 23 million customers in Texas. As a single balancing authority (BA) without synchronous connections to its neighboring systems, ERCOT relies purely on its internal resources to balance power shortages and variations. More than 35 GW of wind generation and 10 GW of solar generation was connected to the ERCOT grid by 2021. This chapter presents a new framework to re-design AS market, which is being implemented at ERCOT in the anticipation of the reliability need arising from a declining system inertia. The salient features of the new AS market include:

1. The new framework considers the balancing need of the grid across a variety of time scales so that the array of AS products will be coordinated holistically when deployed.
2. The new design of AS market is able to accommodate difference resources in providing PFR capability although the characteristics, opportunity costs, economics, and practicality of these resources vary.

This chapter is organized as follows. Existing AS market at ERCOT is reviewed in Sect. 2. Section 3 presents the inertia trend and primary frequency control at ERCOT. The new AS market is discussed in Sect. 4. The details for the new AS product, i.e., fast frequency response (FFR), are provided in Sect. 5. Sections 6 and 7 describe the methods to determine the maximum amount of FFR allowed and the benefits of FFR, respectively. The conclusion is given in Sect. 8.

2 Existing Ancillary Service Market at ERCOT

ERCOT is running a security-constrained economic dispatch (SCED) market every 5 minutes to operate the system at least cost while managing the reliability. In order to manage reliability, SCED must dispatch resources to balance generation with load demand while operating the transmission system within established limits.

However, load and generation are constantly changing, due to daily load patterns, instantaneous load variations, changes in variable generation output, and disconnection of generators. Thus, ASs are procured in the day-ahead market at ERCOT to ensure that the reserve capacity is available in real time to address variability that cannot be covered by the 5-minute energy market. Once the day-ahead award for ASs is cleared, it will become physically binding in real time, i.e., those resources need to fulfill their AS obligations in real time once awarded in the day-ahead market. There is an ongoing effort to change the market design such that the provision of ASs will be co-optimized with the energy in real time after 2024.

An efficient and well-functioning AS market is critical to provide incentives to these qualified resources so that the sufficient resource capacity is reserved which can be deployed in a timely manner to restore the balance between the load and generation. The wholesale market design and rules have been evolving in the past.

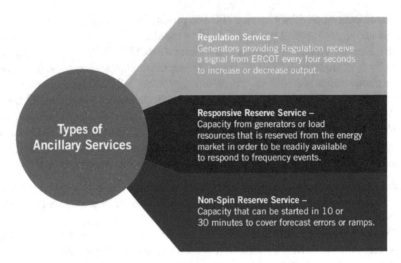

Fig. 6.1 Overview of AS products at ERCOT

Despite this, the AS products procured in the AS market remain unchanged since the inception of the AS market. These AS products, which are primarily characterized by their response time and deployment mechanism, have been introduced to meet North American Electric Reliability Corporation (NERC) requirements. Similar to other AS markets, AS products at ERCOT consist of regulation service (up and down), responsive reserve service (RRS), and non-spin reserve service (NSRS) before the AS market was restructured—see Fig. 6.1. The major characteristics of these AS products are briefly described as follows.

2.1 Regulation Service

Regulation service is needed to correct actual frequency to scheduled frequency and to ensure NERC reliability requirement (BAL-001) is met. Regulation service is provided by the resources that can be deployed every 4 seconds to compensate for the load or generation variations within the SCED time intervals, i.e., ERCOT sends load frequency control (LFC) signal every 4 seconds to increase or decrease power output to the generators providing regulation services. The amount of regulation service needed is determined by the historical usage of regulation service and 5-minute net load changes [32].

2.2   *Responsive Reserve Service (RRS)*

Per NERC standard BAL-003 "Interconnection Frequency Response Obligation," ERCOT must procure a sufficient amount of RRS to avoid the activation of under-frequency load shed (UFLS) at 59.3 Hz for loss of two largest generation units within the ERCOT interconnection.

RRS can be provided by the capacity from generators or load resources that are readily available to respond to frequency excursions during unit trips. Two types of frequency response characteristics exist currently. First, the governors of thermal generating units begin to respond "immediately" but take a few seconds to provide significant deployment (since they require more steam or more combustion). Second, load resources (LRs) providing RRS have under-frequency relays that respond in approximately 0.5 s when frequency drops below 59.7 Hz.

2.3   *Non-spin Reserve Service (NSRS)*

NSRS is provided by those resources that can be started in 10 or 30 minutes to cover net load forecast errors or ramps. These resources consist of generation resources capable of being ramped to a specified output level within 30 minutes and load resources that are capable of being interrupted within 30 minutes and that are capable of running (or being interrupted) at a specified output level for at least 1 hour. NSRS is required to meet NERC BAL-002 (to restore contingency reserve within 90 minutes).

NSRS may be deployed to replace loss of generating capacity, to compensate for net load forecast uncertainty on days in which large amounts of spinning reserve are not available online, to address the risk of net load ramp, or when there is a limited amount of capacity available for SCED. Historically, the need for NSRS has occurred during hot or cold days with unexpected changes in weather or following large unit trips to replenish reserves.

3   Inertia Trend and Primary Frequency Control at ERCOT

As more VERs are integrated to the grid, the system inertia could decline. This section presents an overview of inertia trend and how the primary frequency control is coordinated at ERCOT.

3.1   Inertia Trend at ERCOT

The power system's primary frequency response performance is highly dependent on the system inertia and how the primary frequency control is activated.

Rotating turbine generators and motors that are synchronously connected to the system store kinetic energy. In response to a sudden loss of generation, kinetic energy will automatically be extracted from the rotating synchronous machines causing the machines to slow down as the system frequency is decaying. Inertia response provides an important contribution to reliability in the initial moments following a generation or load trip event and determines the initial rate of change of frequency [9–11].

The amount of the system inertia depends on the number and size of generators and motors synchronized to the grid. It is difficult to account for the contribution of motor loads to the system inertia as the statuses of motors are unknown to system operators [19]. Therefore, inertia response of motor load can be included into load damping constant.

The system inertia from all online synchronous generators is calculated as

$$M_{sys} = \sum_{i \in I} H_i \cdot MVA_i \qquad (6.1)$$

where H_i and MVA_i are the inertia constant and installation MVA capacity of synchronous machine i, respectively, and I is the online synchronous unit set.

The system inertia at ERCOT decreases over the years as shown in Fig. 6.2, which depicts the hourly system inertia and the corresponding wind penetration, i.e., the portion of load supplied by wind generation, between 2015 and 2017. The lowest inertia in each year has dropped from 152 GW·s in 2015 to 130 GW·s in 2017. The

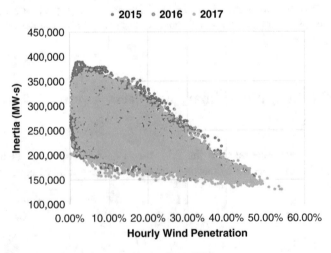

Fig. 6.2 Inertia vs. hourly wind penetration for 2015–2017

decrease in the inertia was partially attributed to the reduction in the net load. If there is an abundant wind or solar generation coinciding with a low load condition, the wholesale energy market prices can be low or even negative. Under these circumstances, synchronous generators may be offline for economic reasons. As a result, a low net load case could lead to fewer generators committed online, then reducing the system inertia.

3.2   Overview of Frequency Control Coordination at ERCOT

Figure 6.3 depicts an overview of how the primary frequency control is coordinated for a severe under-frequency event at ERCOT. When the frequency drops below 59.91 Hz, the RRS capacity carried by the generation units is released to SCED. After a delay of 1 minute, SCED will be re-run with the updated resource limits so that the base points for these RRS generation units will be increased to assist in restoring the frequency. If the frequency continues to drop below 59.8 Hz, the hydro resources operated as synchronous condensers will respond autonomously. If the frequency stays below 59.7 Hz, LRs will be tripped offline to arrest the decline of the frequency. Once the frequency is restored back above 59.98 Hz, the deployment of RRS will be recalled. At ERCOT, the amount of the hydro resources operated as synchronous condensers is relatively small compared to the generation units and LRs in the provision of RRS.

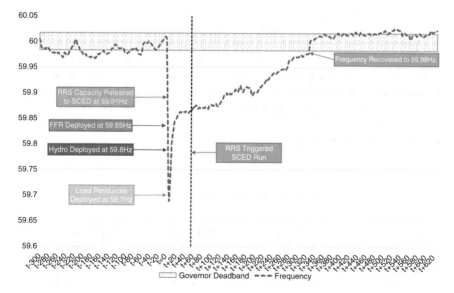

Fig. 6.3 Coordination of primary frequency control at ERCOT

3.3 RRS at ERCOT Before Re-design of AS Market

Having a sufficient amount of RRS is critical to arrest frequency excursions within a few seconds following generation unit trips. Both the generators providing governor response and LRs with under-frequency relays are eligible to participate in RRS market. The following describes their features in the provision of primary frequency control.

When RRS is provided by online synchronous generators through governor response or governor-like actions to arrest frequency deviations, it is termed as a PFR service. As a single BA, ERCOT must comply with the BAL-003 standard. The frequency response obligation for ERCOT is 413 MW/0.1 Hz. To meet this requirement, ERCOT requires every resource with a speed governor to put the governor in service whenever the resource is online. In addition, the droop setting should not exceed 5%, and the frequency response deadband should not be no more than ± 0.018 Hz.

LRs on under-frequency relay (UFR) can also provide RRS if they can be self-deployed to provide a full response within 30 cycles after the frequency meets or drops below a certain threshold (59.7 Hz). LRs on UFR are equipped with an under-frequency relay to arrest the quick frequency decline. As required by ERCOT, the response time for LRs should be less than 500 ms (including the frequency relay pickup delay and the breaker action time). This makes the response of LRs more effective to mitigate the decline of frequency compared to the generators because a generator needs a few seconds to react to the change in the frequency to provide the primary frequency response. Therefore, the deployment of LRs is able to improve the frequency nadir and is instrumental in preventing frequency from dropping below the involuntary UFLS threshold when losing large generation units.

Historically, ERCOT has relied more on PFR and LRs to protect the grid against the large disturbances. Nowadays, there is an increasing interest in deploying energy storage resources in a large scale in the future power grid. This opens up a new opportunity to utilize these fast-acting resources to enhance the primary frequency control performance. ERCOT is in the process of developing new market rules to allow batteries to respond at 59.85 Hz to provide a FFR reserve, which is described in details in the next section.

4   New Ancillary Service Market

As the system inertia is declining, fast frequency response can significantly improve the frequency control performance. On the other hand, the past market rules at ERCOT required that RRS should be deployed within 10 minutes upon the receipt of the deployment instructions. This requirement does not explicitly incentivize the resources which can quickly deliver the primary frequency response before the frequency nadir (point C). In addition, this could create unnecessary barriers for emerging technologies like batteries to participate in the RRS market. To this end, it is imperative to re-design the AS market, especially RRS, in order to improve both economic and reliable operations for a future power grid with a large amount of VERs.

The fundamental principles to guide the design of such new AS market are given as follows.

1. To meet NERC reliability requirement so as to maintain a satisfactory frequency control performance
2. To allow batteries to be awarded FFR to improve the efficiency of the AS market and reduce the operational cost
3. To accommodate the characteristics of a variety of new market entrants
4. To coordinate the deployment actions of AS products at different time scales

Based on these principles, a new AS market framework has been proposed and is being implemented at ERCOT, which consists of four AS products, i.e., regulation service, RRS (new), ERCOT contingency reserve service (ECRS), and NSRS—see Fig. 6.4.

The key factor distinguishing these new AS products is still the response time required for each reserve. Regarding regulation service, there is no change proposed to its qualifications and deployment mechanism.

Within this new framework, RRS (old) will be unbundled into two products: RRS (new) and ECRS. RRS (new) includes three subsets: PFR, LRs, and FFR. All of three RRS subsets are procured with an intention to arrest large frequency excursions following a generation trip. More detailed description of FFR will be presented in Sect. 5.

ECRS is a new AS product introduced to restore RRS (new) responsibility once RRS resources are depleted or to mitigate a reliability concern if there is a deficiency

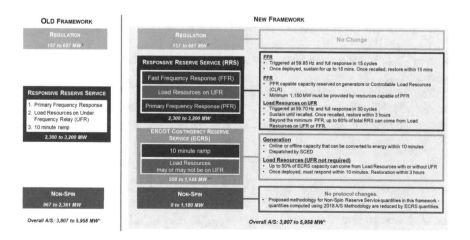

Fig. 6.4 New AS products at ERCOT

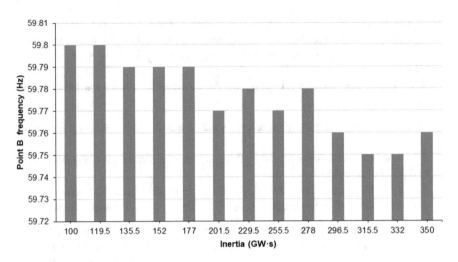

Fig. 6.5 Point B vs. inertia as a result of a single largest unit outage

in the ramping capacity. By design, ECRS can be dispatched by SCED and should respond within 10 minutes to the deployment instructions.

ECRS is required to meet NERC BAL-002 (to recover frequency within 15 minutes). The minimum amount of ECRS is determined in two steps. In the first step, dynamic simulations are performed to identify the frequency response (point B) when a single largest unit outage causes the frequency nadir (point C) to drop at 59.7 Hz (LRs will not be tripped in this case). Figure 6.5 shows different point B as a function of the system inertia.

In the second step, to fulfill the obligation of restoring point B frequency to above 59.98 Hz within the 10 minutes, the amount of ECRS required is calculated by

$$p_{\text{ECRS}} = \left(59.98 - f_{B\,\text{Point}}^{M_{\text{sys}}}\right) \cdot p_{\text{Load}} \cdot \frac{10}{100} \tag{6.2}$$

where $f_{B\,\text{Point}}^{M_{\text{sys}}}$ is the B point frequency for the system inertia (M_{sys}) and p_{Load} is the load demand.

The qualification criterion and the response time for NSRS will remain unchanged. However, the minimum amount of NSRS required will be modified. Before the new AS design is in effect, the amount of NSRS needed is given by

$$p_{\text{NSRS}} = \Omega(\varepsilon_{\text{load,3h}} - \varepsilon_{\text{wind,3h}} - \varepsilon_{\text{solar,3h}}) \tag{6.3}$$

where $\varepsilon_{\text{load, 3h}}$, $\varepsilon_{\text{wind, 3h}}$, and $\varepsilon_{\text{solar, 3h}}$ are the 3-hour-ahead load, wind, and solar generation forecast error, respectively, and Ω is the percentile function applied to the net load forecast error data set ($\varepsilon_{\text{load, 3h}} - \varepsilon_{\text{wind, 3h}} - \varepsilon_{\text{solar, 3h}}$) [33].

After the new AS design is implemented, the amount of NSRS needed will be

$$p_{\text{NSRS}} = \max\left(0, \Omega(\varepsilon_{\text{load,3h}} - \varepsilon_{\text{wind,3h}} - \varepsilon_{\text{solar,3h}}) - p_{\text{ECRS}}\right) \qquad (6.4)$$

Equation (6.3) represents the way how NSRS was determined before ECRS is introduced. Since ECRS is a 10-minute AS product, it can substitute NSRS (30-minute AS product) in order to meet NERC BAL-002 requirement. Therefore, the minimum amount of NSRS needed will be significantly reduced after the introduction of ECRS as indicated by (6.4).

5 Fast Frequency Response (FFR)

To be qualified for the provision of FFR, a resource should be able to be automatically deployed and provide its full response within 15 cycles after the frequency meets or drops below a preset threshold (59.85 Hz) or be deployed via a verbal dispatch instruction (VDI) within 10 minutes. FFR resources must sustain a full response for at least 15 minutes once deployed. When a resource providing RRS as FFR is deployed, it shall not recall its capacity until the system frequency is greater than 59.98 Hz or they have been sustainably deployed over 15 minutes. Once recalled, the resources providing FFR must restore their full FFR responsibility within 15 minutes after the cessation of deployment or as otherwise directed by ERCOT.

Table 6.1 shows a comparison of key features between FFR and LRs. In comparison to LRs, FFR will be deployed earlier and faster when a loss of generation event happens. Earlier response from FFR will aid in preserving LRs providing RRS for more severe events. With a trigger set at 59.85 Hz, FFR will deploy more frequently than LRs in the response to the under-frequency events. As FFR resources provide a frequency response, most frequency events will not trigger the LR deployment, thus preserving the frequency response capabilities from LRs to be available for the next severe event as shown.

Shorter restoration time for FFR resources also limits ERCOT's exposure (i.e., inability to respond) to the next event with a similar magnitude. FFR resources upon the deployment completion will reset themselves and become available to respond to another event within 15 minutes. In contrast, LRs providing RRS are allowed to reset themselves with 3 hours after a RRS deployment and become available to respond to

Table 6.1 Comparison between FFR and LR

	FFR	LR
Frequency trigger	59.85 Hz	59.7 Hz
Response time	15 cycles	30 cycles
Sustained deployment period	15 minutes	1 hour
Restoration time	15 minutes	3 hours
Deployment during EEAs	VDI	VDI

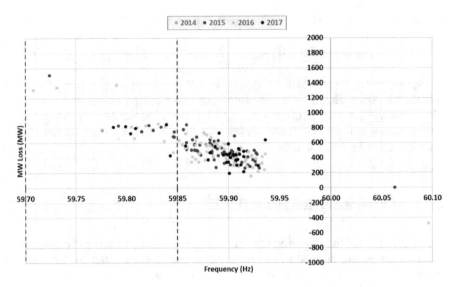

Fig. 6.6 MW loss vs. frequency nadir at ERCOT

another event. If a frequency event triggers the LR providing RRS to be deployed, for the 3-hour duration following such an event, ERCOT may not have adequate frequency responsive resources to respond to another large disturbance.

During energy emergency alert (EEA) conditions, FFR resources may be deployed through VDI. As required, FFR resource should not withdraw energy from the grid until the EEA event has ended.

The response time and the frequency trigger are two key design parameters for the FFR reserve, which not only impact the overall frequency control performance but also influence the market liquidity and efficiency. The determination of these two parameters should consider the performance trend of the ERCOT grid to ensure that FFR resources are effective in protecting the system frequency. In addition, the capabilities of the new technologies should be taken into account in order to attract as many participants in the RRS market as possible. Figure 6.6 shows the historical trend of ERCOT system described by the magnitude of the power losses and the corresponding frequency nadir point. A selection of the frequency trigger at 59.85 Hz can protect against the large disturbances without a need to activate LRs while avoiding the unintended responses to numerous insignificant events. The response time required for FFR is 15 cycles, which is attainable by fast-acting energy storage resources or fast protection relays/breakers.

5.1 Qualifications of FFR and Performance Evaluations

ERCOT will qualify resources that can provide FFR via a qualifications test. A resource must meet the following two requirements in order to pass the test:

1. A resource must respond within 10 minutes of receiving test instruction.
2. A resource's response must be within 95% and 110% of the minimum between the RRS obligation and the maximum deployment allowed to respond.[1]

ERCOT may revoke a resource's FFR qualification if it has two performance failures during actual deployments (manual or frequency triggered) within a rolling 365-day period. The actual performance of a FFR resource is evaluated using the following metric:

- A FFR resource must be deployed in 15 cycles (or 10 minutes for verbal deployment) after the frequency reaches the trip threshold.
- A resource must sustain the response for at least 15 minutes or till ERCOT recalls deployment, whichever occurs first.
- A resource must be reset and made available for the next event within 15 minutes after the deployment is ended.

For each FFR deployment event, the following data will be collected for evaluating FFR response performance:

- High-speed event data from the FFR resources that are not deployed via breaker action
- High-speed event data from the recorders at the primary and backup facilities
- High-speed event data from phasor measurement units available to ERCOT
- Telemetry data for all resources providing FFR during the event
- Recording of frequency and power output with a resolution of no less than 30 samples per second

5.2 Telemetry Data Requirement for Deployment and Recall of FFR

The FFR resources respond autonomously to the local frequency excursions so that no centralized mechanism is needed to activate the deployment of the FFR. However, certain coordination is necessary when the deployment of FFR resources is recalled in order not to harm the frequency recovery. One example of telemetry data for a FFR resource is given in Fig. 6.7. Once the frequency is below 59.85 Hz,

---

[1]The maximum deployment allowed to respond equals to telemetered high sustained limit (HSL)-low sustained limit (LSL) for modeled generation resource (generators and batteries) or maximum power consumption (MPC)-low power consumption (LPC) for modeled load resource.

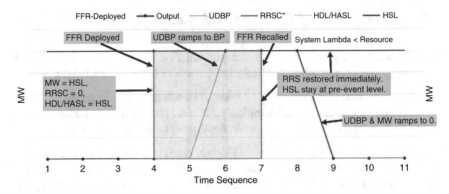

Fig. 6.7 Telemetry example for deployment and recall of FFR (for illustration only) (UDBP updated desired base point, RRSC RRS schedule, HSL high sustainable limit, HDL high dispatch limit, HASL high ancillary service limit)

ERCOT will send the FFR resource the deployment instruction so that the FFR resource will update its RRS schedule (RRSC) and inform ERCOT the new RRSC, which makes the reserved FFR capacity available to SCED. After the execution of the next SCED is completed, ERCOT will provide the FFR resource a new updated desired base point (UDBP), ramping from pre-event level to the full deployment. However, the actual FFR deployment has occurred prior to this so that the FFR resource is not required to follow UDBP when deployed. After the cessation of 15 minutes, the FFR resource is recalled so that its HSL telemetry data returns back to pre-event level. The next SCED run will update the UDBP, ramping down to 0 MW, which the FFR resource must follow. In this way, a FFR resource gradually and smoothly reduces its output to 0 MW once recalled.

6   Maximum Amount of FFR Allowed

The deployment of FFR could quickly arrest the frequency decline. One example is given in Fig. 6.8, which simulates the frequency response of the ERCOT grid when a generation producing a 947 MW of power output is tripped offline. A full-detailed ERCOT dynamic model is used in the dynamic simulation, which is composed of over 500 generators and 10,000 buses. For this particular disturbance, 525 MW of FFR resources were deployed, and as a result, the frequency nadir has been prevented from dropping below 59.7 Hz as shown in Fig. 6.9. This clearly demonstrates the effectiveness of FFR in avoiding the activation of the first-stage UFLS.

However, a reliability concern could arise if an excessive amount of FFR resources is deployed, leading to a potential frequency overshoot. Figure 6.8 depicts the overshoot of frequency due to activation of different amount of FFR when the system inertia is 100 GW·s (100 GW·s is the minimum system inertia requirement for ERCOT). In this case, 292 MW of generation was lost at 0.5 second, causing the

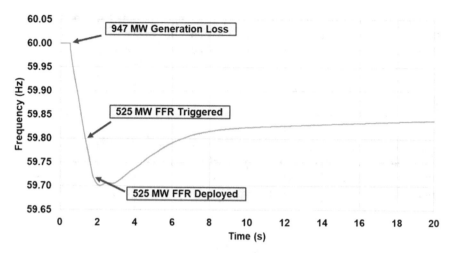

Fig. 6.8 Simulated frequency response for a 947 MW of generation loss

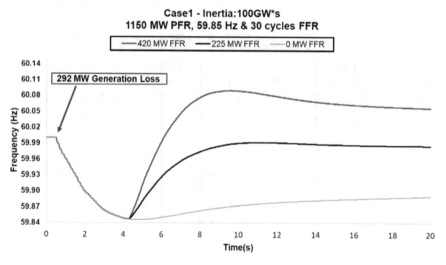

Fig. 6.9 Frequency overshoot due to the deployment of FFR (red, 420 MW FFR; black, 225 MW FFR; blue, 0 MW FFR)

frequency to drop below 59.85 Hz. If 420 MW of FFR has responded to this, the frequency will recover more quickly, but the highest frequency can reach as high as 60.10 Hz. In the case when an excessive amount of FFR reacts to this event, this post-disturbance over-frequency response may cause another need to deploy regulation down resources in order to bring the system frequency back to 60 Hz.

However, when the system inertia is extremely low, over-generation may likely happen, i.e., thermal units may have to be operated close to their minimum output limit and thus have a limited capability of reducing their power production in the response to an over-frequency event. To mitigate this concern, ERCOT currently limits the maximum system-wide FFR responsibility to 420 MW.

7 Benefits of FFR

The earlier and faster response of FFR can benefit the grid's reliability and operational efficiency. In particular, a new approach is presented in this section, which can qualify the benefits of FFR to the grid operations in two terms: (1) the reduction in critical inertia and (2) RRS cost saving.

7.1 Impact of FFR over Critical Inertia

In a low inertia grid like ERCOT, the grid frequency could decline very quickly following a resource trip. LRs begin to react once the frequency drops below their triggering frequency, f_{LR} (59.7 Hz). However, it takes a delay, d_{LR} (30 cycles), for LRs to completely open their breaker. During this period of d_{LR}, the system frequency will continue to decrease. If the rate of change of the frequency is very high, there could be a possibility that the frequency can drop below the trigger of first-stage UFLS (59.3 Hz) before the breaker of LRs can be fully opened. This scenario can be depicted in Fig. 6.10 and be mathematically described as

$$t_{LR} + d_{LR} > t_{UFLS,1st} \tag{6.5}$$

Fig. 6.10 Frequency response when $t_{LR} + d_{LR} > t_{UFLS,\ 1st}$ (t_{LR} and $t_{UFLS,\ 1st}$ are when the frequency reaches 59.7 Hz and 59.3 Hz, respectively)

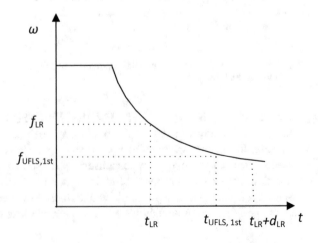

To prevent this from happening, LRs need to have a sufficient amount of time to respond to the frequency excursions. One remedy to this is to maintain a minimum amount of inertia (critical inertia), M_{min}, on the grid, which ensures that the condition $t_{LR} + d_{LR} < t_{UFLS,1st}$ is satisfied. When the system inertia is above this minimum inertia level, M_{min}, LRs are still effectively tripped offline to arrest the frequency drop. Otherwise, the frequency will drop below the first-stage UFLS triggering frequency before LRs can provide sufficient frequency response, following the worst contingency, i.e., the simultaneous loss of two largest online generators.

Through the dynamic simulation, the critical inertia at ERCOT is estimated to be 100 GW·s when FFR resources are not considered.[2] With the appropriately selected trip settings stated in Sect. 5, FFR can help to reduce this critical inertia. If 420 MW of FFR can be fully deployed with 15 cycles, it can reduce the critical inertia from 100 GW·s to 88 GW·s. This is because an earlier and faster response from FFR can decrease the rate of change of the frequency and thus provide LRs more time to deliver their full response. A reduction in the critical inertia can facilitate more renewable resources to be reliably integrated in the ERCOT grid. Otherwise, thermal units have to be uneconomically committed to maintain the system inertia above the critical inertia even though the abundant energy could be produced from wind or solar renewable resources.

7.2 RRS Cost Saving with FFR Resources

RRS is procured to ensure that the sufficient capacity is available to respond to frequency excursions once a unit trips. Prior to the next operational year, ERCOT sets the minimum RRS requirement for the expected grid operations, which varies by hour of the day and by month [31]. One key factor influencing how much RRS is needed is the system inertia. This dependence is considered here when deciding the need for RRS. The basic approach to determine RRS requirement for each hour of the next year consists of two steps: (1) to project the system inertia conditions for the next year by creating a time-series data for 8760 hours and (2) to map from the projected inertia condition to the RRS need. This two-step approach is briefly described as follows, and more details can be found in [31].

The first step is to calculate future inertia conditions for each 4-hour interval of the next year. This is based on expected diurnal load and wind/solar patterns for the same hour block of the same month in the past 2 years. A percentile is then applied to this data set to produce a future inertia condition for each 4-hour block of the same month for the next year.

---

[2]Additionally, simulated conditions show the wide-area voltage oscillations at inertia below 100 GW·s.

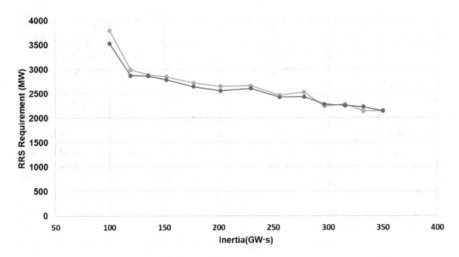

Fig. 6.11 Reduction in RRS requirement due to the inclusion of FFR (blue, PFR + LR; black, PFR + LR + FFR)

The second step is to determine the minimum amount of RRS requirement based on the future 8760-hour inertia conditions, through a look-up table. Without including FFR, the RRS need corresponding to an inertia level for 12 different cases is derived from dynamics studies in which the loss of 2 largest online units is simulated—see the blue line in Fig. 6.10. It can be seen that the RRS need decreases dramatically as the system inertia increases. This is because when there are more generators online under heavy loading conditions, i.e., the overall system inertia is higher, the rate of change of frequency following the disturbance is much smaller than in a low inertia condition. As a result, the aggregated slow-acting governor-like PFR response has enough time to react to the loss of generation.

After 420 MW of FFR resources are used, the total RRS requirement will be reduced, as shown in the black line in Fig. 6.11. This decrease in the RRS requirement is larger at the low system inertia levels and becomes diminished as the system inertia increases.

In order to evaluate the potential benefits of FFR, it is estimated that if the new AS market were in effect in 2019, and FFR offers of 420 MW were procured in each hour, there may be a lower quantity of RRS procured in 7272 hours out of 8016 hours for January through November 2019. The reduction in RRS quantities would vary between 0 MW and 81 MW during these 8016 hours resulting in an overall reduction during this period of 244,712 MWh in required RRS quantities.

The historical RRS market clearing prices for capacity (MCPCs) from January 1, 2018, through November 30, 2018, are used as a proxy to estimate the magnitude of the financial impact of the decrease in the minimum RRS requirements. Assuming 244,712 MWh in required RRS quantities for January through November 2019, the reduction in total RRS quantity by hour is multiplied by 2018 hourly MCPCs for RRS results in an estimated savings of $3,426,088. The estimated cost saving is expected to increase in the future as ERCOT experiences higher frequency of lower inertia periods and procures higher quantities of RRS.[3]

8 Conclusions

AS market awards those services critical to maintain the reliability and security of the power systems through a market mechanism. Currently, the power industry is undergoing a major transformation. The increasing penetration of VERs will dramatically change the reliability need of a future power grid and thus calls for re-design of AS market. On the other hand, it is desirable to achieve a balance between reliability and economics while opening up the AS market to both traditional and non-traditional market participants.

This chapter presents a new AS framework being implemented at ERCOT in order to address the primary frequency control issues associated with the declining system inertia. The essential innovation is to divide RRS into three sub-categories (PFR, LRs on UFR, and FFR) so that it explicitly awards those resources which can be quickly deployed to arrest the frequency drop following the loss of the large online generation units. To meet NERC BAL-002, ECRS is also introduced to recover frequency within 15 minutes. This new design makes it suitable for energy-limited resources like batteries to offer FFR in the AS market.

In addition, the detailed description of designing key performance metrics and qualifying the benefits of FFR was provided in the chapter. It was found that the earlier and faster response of FFR can bring significant benefits to the grid's reliability. Introduction of FFR can reduce the critical inertia from 100 GW·s to 88 GW·s, and the estimated cost saving amounts to $3,426,088 if the new design were implemented for January through November 2019. On the other hand, if an excessive amount of FFR resources is deployed, it could lead to a potential frequency overshoot. Therefore, the maximum system-wide FFR responsibility at ERCOT is limited to 420 MW.

The experiences gained at ERCOT can be beneficial to other regions which also face the similar challenges when dealing with a high penetration of VERs. Future work will focus on the study and analysis of the performance of FFR and its impact over the market operations and grid reliability.

---

[3] In addition, with an increase in the quantity of resources qualified to provide RRS and able to submit offers for RRS, it is likely the clearing price of RRS will decrease.

References

1. Promoting Wholesale Competition Through Open Access Nondiscriminatory Transmission Services by Public Utilities, Order No. 888, FERC, 1996
2. Mohammad Shahidehpour, Hatim Yamin, and Zuyi Li, Market Operations in Electric Power Systems: Forecasting, Scheduling, and Risk Management, John Wiley & Sons, 2003.
3. Litvinov, Eugene, Feng Zhao, and Tongxin Zheng, "Electricity markets in the United States: Power industry restructuring processes for the present and future," IEEE Power and Energy Magazine, vol.17, no. 1, 2019, pp. 32-42.
4. Yann G. Rebours, Daniel S. Kirschen, Marc Trotignon, and Sbastien Rossignol, "A survey of frequency and voltage control ancillary services—Part I: Technical features," IEEE Transactions on power systems, vol. 22, no. 1, 2007, pp. 350-357.
5. E. Ela, and A. Tuohy, "The evolution of ancillary services to facilitate integration of variable renewable energy resources: A survey of some changes to the ancillary services and ancillary service markets," Electric Power Res. Inst., Tech. Update# 3002008987, 2016.
6. Zhi Zhou, Todd Levin, and Guenter Conzelman, Survey of US Ancillary Services Markets, Argonne National Lab (ANL), 2016.
7. Pengwei Du, Ross Baldick and Aidan Tuohy, Integration of Large-Scale Renewable Energy into Bulk Power Systems, Springer International Publishing, 2017.
8. Julia Matevosyan, Pengwei Du, "Wind generation scheduling and coordination in ERCOT nodal market," in Integration of Large Scale Renewable Energy into Bulk Power System: From Planning to Operation, 2017
9. P. Kundur, Power System Stability and Control, McGraw-Hill, 1994.
10. Joseph H Eto, "Use of frequency response metrics to assess the planning and operating requirements for reliable integration of variable renewable generation," Lawrence Berkeley National Laboratory, 2011.
11. NERC Frequency Response Initiative, April, 2010.
12. Mohammad Dreidy, H. Mokhlis, and Saad Mekhilef, "Inertia response and frequency control techniques for renewable energy sources: A review," Renewable and sustainable energy reviews, no. 69, 2017, pp: 144-155.
13. Essential Reliability Services and the Evolving Bulk-Power System—Primary Frequency Response, FERC Order 842, February 15, 2018.
14. Jon O'Sullivan, Alan Rogers, Damian Flynn, Paul Smith, Alan Mullane, and Mark O'Malley, "Studying the maximum instantaneous non-synchronous generation in an Island system—Frequency stability challenges in Ireland," IEEE Transactions on Power Systems, vol.29, no.6, 2014, pp. 2943-2951.
15. Ahmad Shabir Ahmadyar, Shariq Riaz, Gregor Verbič, Archie Chapman, and David J. Hill, "A framework for assessing renewable integration limits with respect to frequency performance," IEEE Transactions on Power Systems, vol.33, no.4, 2018, pp. 4444-4453.
16. Mengran Yu, "Instantaneous penetration level limits of non-synchronous devices in the British power system," IET Renewable Power Generation, vol.11, no.8, 2016, pp. 1211-1217.
17. Yong Liu, Shutang You, Jin Tan, Yingchen Zhang, and Yilu Liu. "Frequency response assessment and enhancement of the U.S. power grids toward extra-high photovoltaic generation penetrations—an industry perspective," IEEE Transactions on Power Systems, vol.33, no.3, 2018, pp. 3438-3449.
18. Shutang You, Yong Liu, Jin Tan, Melanie T. Gonzalez, Xuemeng Zhang, Yingchen Zhang, and Yilu Liu, "Comparative assessment of tactics to improve primary frequency response without curtailing solar output in high photovoltaic interconnection grids," IEEE Transactions on Sustainable Energy, vol.10, no.2, 2018, pp.718-728.
19. Pengwei Du and Julia Matevosyan, "Forecast system inertia condition and its impact to integrate more renewables," IEEE Transactions on Smart Grid, vol.9, no.2, 2018, pp. 1531-1533.

20. Ronan Doherty, Alan Mullane, Gillian (Lalor) Nolan, Daniel J. Burke, Alexander Bryson, and Mark O'Malley, "An assessment of the impact of wind generation on system frequency control," IEEE Transactions on Power Systems, vol.25, no.1, 2010, pp. 452-460.
21. Ti Xu, Wonhyeok Jang, and Thomas Overbye, "Commitment of fast-responding storage devices to mimic inertia for the enhancement of primary frequency response," IEEE Transactions on Power Systems, vol.33, no.2, 2018, pp. 1219-1230.
22. Uthier Delille, Bruno Francois, and Gilles Malarange, "Dynamic frequency control support by energy storage to reduce the impact of wind and solar generation on isolated power system's inertia," IEEE Transactions on Sustainable Energy, vol.3, no.4, 2012, pp. 931-939.
23. Angel Molina-García, François Bouffard, Daniel S. Kirschen, "Decentralized demand-side contribution to primary frequency control," IEEE Transactions on Power Systems, vol.26, no.1, Feb. 2011, pp. 411-419.
24. Ronan Doherty, Gillian Lalor, and Mark O'Malley, "Frequency control in competitive electricity market dispatch," IEEE Transactions on Power Systems, vol.20, no.3, 2005, pp. 1588-1596.
25. Erik Ela, Vahan Gevorgian, Aidan Tuohy, Brendan Kirby, Michael Milligan, and Mark O'Malley, "Market designs for the primary frequency response ancillary service—part I: motivation and design," IEEE Transactions on Power Systems, vol.29, no.1, Jan. 2014, pp. 421 – 431.
26. Erik Ela, Vahan Gevorgian, Aidan Tuohy, Brendan Kirby, Michael Milligan, and Mark O'Malley, "Market designs for the primary frequency response ancillary service—part II: case studies," IEEE Transactions on Power Systems, vol.29, no.1, Jan. 2014, pp. 432 – 440.
27. Weifeng Li, Pengwei Du, Ning Lu, "PFR ancillary service in low-inertia power system," IET Generation, Transmission & Distribution, 2020.
28. Weifeng Li, Pengwei Du, and Ning Lu, "Design of a new primary frequency control market for hosting frequency response reserve offers from both generators and loads," IEEE Transactions on Smart Grid, vol.9, no.5, 2018, pp. 4883-4892.
29. Cong Liu and Pengwei Du, "Participation of load resources in day-ahead market to provide primary-frequency response reserve," IEEE Transactions on Power Systems, vol.33, no.5, 2018, pp. 5041-5051.
30. N. Miller, D. Lew, and R. Piwko. "Technology capabilities for fast frequency response," GE Energy Consulting, Tech. Report, 2017.
31. Pengwei Du and Weifeng Li. "Frequency Response Impact of Integration of HVDC into a Low-inertia AC Power Grid," IEEE Transactions on Power Systems, 2020.
32. Pengwei Du, Hailong Hui, and Ning Lu, "Procurement of regulation services for a grid with high-penetration wind generation resources: a case study of ERCOT," IET Generation, Transmission & Distribution, vol. 10, no. 16, 2016, pp. 4085-4093.
33. Nischal Rajbhandari, Weifeng Li, Pengwei Du, Sandip Sharma, and Bill Blevins, "Analysis of net-load forecast error and new methodology to determine non-spin reserve service requirement," In 2016 IEEE Power and Energy Society General Meeting (PESGM), pp. 1-5, IEEE, 2016.

Chapter 7
System Inertia Trend and Critical Inertia

1 Basics of Synchronous Inertia

Rotating turbine generators and motors that are synchronously connected to the system store kinetic energy [1–3]. In response to a sudden loss of generation, kinetic energy will automatically be extracted from the rotating synchronous machines causing the machines to slow down, and, as a consequence, system frequency is decaying. This is called inertial response. Inertial response provides an important contribution to reliability in the initial moments following a generation or load trip event determining the initial rate of change of frequency.

The amount of inertia depends on the number and size of generators and motors synchronized to the system. It may be difficult to account for motor loads as this information oftentimes is not available to the system operator; therefore, inertial response of motor load is usually lamped into load damping constant discussed later in this chapter.

For any hour, synchronous inertial response (SIR) from the generators is calculated as [4]

$$M_{\text{sys}} = \sum_{i \in I} H_i \cdot \text{MVA}_i \qquad (7.1)$$

where I is the set of online synchronous generators or condensers, MVA_i is the MVA rating of online synchronous generator or synchronous condenser i, and H_i is the inertia constant for online generator or synchronous condenser i in a system (in seconds on machine MVA rating, MVA_i).

Table 7.1 shows the averaged (by unit size) inertia constant for different unit types in ERCOT. By comparison, hydro, simple cycle combustion turbine, gas steam, and coal have smaller inertia constant than nuclear and combined cycle, while non-synchronous resources like wind and solar generation do not contribute to the synchronous inertia.

© The Author(s), under exclusive license to Springer Nature Switzerland AG 2023
P. Du, *Renewable Energy Integration for Bulk Power Systems*, Power Electronics
and Power Systems, https://doi.org/10.1007/978-3-031-28639-1_7

Table 7.1 Averaged inertia constant (H) by fuel type

	Average MVA	Average H (seconds) on average MVA base	Average inertial response contribution H·MVA
Nuclear	1457	4.07	5930
Coal	730	2.7	1971
Combustion turbine	90	54.84	436
Gas steam	266	2.942.88	766
Combined cycle	478	4.89	2337
Hydro	21	2.46	49
Wind	116	0	0
Solar PV	27	0	0

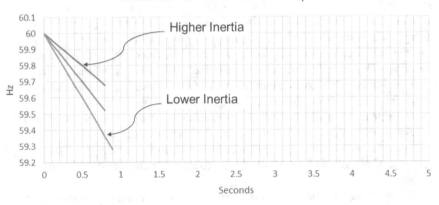

Fig. 7.1 Initial RoCoF after the same unit trip at different inertia conditions

System inertia determines the rate of change of frequency (RoCoF) after an event (generation trip, step change in load). If the system inertia is low, the rate of change of frequency after an event is higher (Fig. 7.1), giving less time for primary frequency response to be fully deployed and arrest frequency decay before the frequency reaches under-frequency load shedding set points, resulting in involuntarily load disconnection.

2   Inertia at ERCOT

The system inertia varies with the hours of day and seasons of year as the statuses of the synchronous generators change over the time. Each time when more synchronous generators are committed (or disconnected from the grid), the system inertia

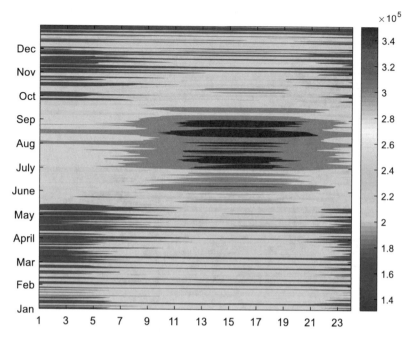

Fig. 7.2 Inertia in 2020

becomes larger (or smaller). Figure 7.2 shows the fluctuating inertia values for 8760 hours in 2020, and the general trend is that the system inertia tends to be high in the summer and low in the shoulder months. This is attributed to the fact that more synchronous generators were online to serve the high load demand in the summer. The lowest inertia in 2020 was 131,122 MW·s, which occurred at 2 a.m. on May 1, 2020.

The system inertia is also correlated with the net load (which is the load minus the aggregated wind and solar generation). If there is an abundant wind or solar generation contemporaneous with low load conditions, wholesale energy market prices can be low or even negative. During these conditions, synchronous generators may be offline for economic reasons, which reduces system inertia. The scatter plot in Fig. 7.3 depicts the hourly inertia conditions for 2015, 2016, and 2017 and corresponding hourly wind penetration, i.e., portion of load supplied by wind generation.

Note that system inertia at any given moment is also affected by wind/solar and load conditions expected in the near future (next few hours) and startup times of the generators. This is because generators with longer startup times would need to start earlier in the anticipation of net load pickup.

Different fuel type units have different commitment patterns due to their characteristics (units' startup times, startup costs, minimum up- and downtimes, maintenance costs, unit production costs, etc.). These commitment patterns in turn effect total system inertia at ERCOT. Figure 7.4 shows one example of the inertia provided

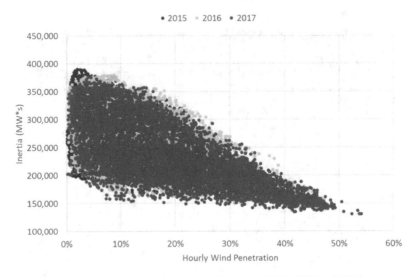

Fig. 7.3 Correlation between wind penetration and inertia in 2015, 2016, and 2017

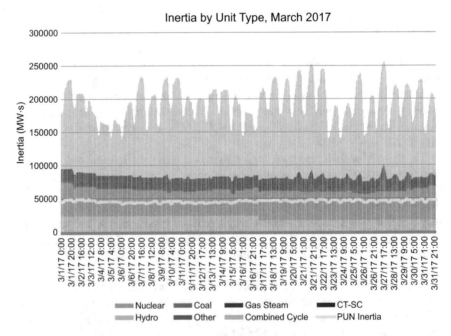

Fig. 7.4 Inertia contribution by fuel type in March 2017; note Private Use Networks (PUN) inertia (shown by the yellow line) is also included in individual unit type inertia (A PUN is an electric network connected to ERCOT transmission network that contains load that is not directly metered by ERCOT)

by nuclear, coal, simple cycle, and combined cycle units in March 2017 in ERCOT. Since the nuclear and coal units are operated as base load units, total contribution from these unit types to system inertia does not vary dramatically over time; however, total system inertial contribution from combined cycle units fluctuates more. Combined cycle units also contribute most to the system inertia (due to high share of combined cycle units in ERCOT resource mix and higher inertial contribution (per MW) compared to other unit types of the same size), while the contribution from simple cycle units is the least (these are more expensive peaking units, and their individual inertial contribution is relatively low).

3 Historical Synchronous Inertia Trends

Figure 7.5 shows the boxplot showing system synchronous inertia for the past several years (2013–2020).[1] The corresponding lowest inertia in each year is given in Table 7.2. The circle on each boxplot is showing inertia during time when the highest portion of load was served by wind generation in that year. In 2019, it was on November 26 3:52 a.m., 57.87%. Some observations can be drawn as follows.

- ERCOT tripled installed wind capacity between 2013 and 2020, adding nearly 20 GW of wind generation (including all synchronized projects).
- An equal amount of coal capacity has about 55 GW·s of inertia contribution.
- Counterintuitively, the inertia trend is not declining as installed wind/solar capacity increases, as was anticipated, due to other factors, such as load growth, gas prices, wind curtailments, etc.

The minimum as well as median (indicated by the red line in each box plot) of the inertia at ERCOT is trending up between 2013 and 2020 (Fig. 7.6), even though the installed capacity of wind generation is increasing over the past few years; see Fig. 7.6.

This can be explained by changes in unit commitment patterns. With lower gas prices in 2014–2015 in general, it was cheaper to run a combined cycle unit than a coal unit. As shown in Figs. 7.7 and 7.8, the inertia contribution from coal units dropped since 2014, and this drop was offset by the increase in the inertia provided by combined cycle. The gas steam and combustion turbine simple cycle units only account for a small portion of the system inertia. The inertia provided by private use network (PUN) units (generation at industrial sites) (Fig. 7.9) has increased since 2014. PUN is mainly composed of combined cycle, combustion turbine simple cycle, and gas steam units.

---

[1] For years 2013–2020, inertia calculation unit status is based on individual unit production (if unit production is higher than 5 MW threshold, a unit is considered online). In 2017, inertia network model-driven inertia calculation was implemented into ERCOT EMS system. This calculation is using unit status directly. Inertia contribution of each online unit is calculated as its inertia constant in seconds multiplied by corresponding MVA base.

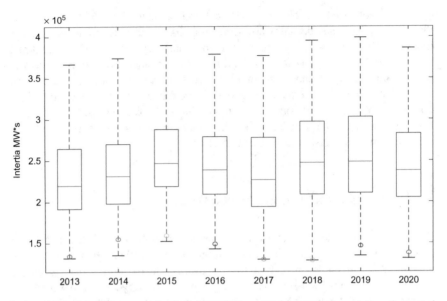

Fig. 7.5 Boxplot of the system inertia in ERCOT from 2013 to 2020 (On each box, the central mark (red line) is the median, the edges of the box (in blue) are the 25th and 75th percentiles, the whiskers correspond to +/− 2.7 sigma (i.e., represent 99.3% coverage, assuming the data are normally distributed), and the outliers are plotted individually (red crosses). If necessary, the whiskers can be adjusted to show a different coverage)

Table 7.2 Lowest inertia in different years (GW·s)

	2013 3/10 3: 00 a.m.	2014 3/30 3: 00 a.m.	2015 11/25 2: 00 a.m.	2016 4/10 2: 00 a.m.	2017 10/27 4: 00 a.m.	2018 11/03 3: 30 a.m.	2019 3/27 1: 00 a.m.	2020 05/01 2: 00 a.m.
Date and time								
Min synch. Inertia (GW·s)	132	135	152	143	130	128.8	134.5	131.1
System load at minimum synch. Inertia (MW)	24,726	24,540	27,190	27,831	28,425	28,397	29,883	30,679
Non-synch. Gen. In % of system load	31	34	42	47	54	53	50	57

Based on 2013–2020 data, PUN inertia contribution has never fallen below 31.5 GW·s and is trending up. Steady system inertia level provided by PUNs can be explained by steam demand of the PUN sites, while upward trend can be explained by lower gas prices.

A breakdown of generation power and inertia at the annual minimum system inertia hours from 2013 to 2020 (Fig. 7.10) also indicates the similar trend as in Figs. 7.7, 7.8, and 7.9. As the contribution from coal units was diminishing, this was offset by the increase in the inertia contribution from combined cycle and simple

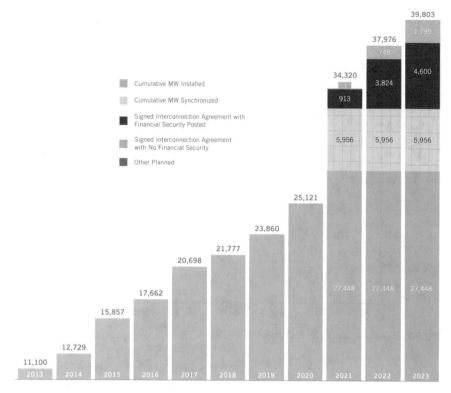

Fig. 7.6 ERCOT wind installations by year (Cumulative MW Synchronized pertains to projects that ERCOT has approved to generate energy for the grid but have not passed all qualification testing necessary to be approved for participation in ERCOT market operations)

cycle units. At these hours, PUNs were generating between 1200 and 6300 MW, and inertial contribution from PUNs was 38–48 GW·s (increasing from 2013 to 2020).

4 Determining Critical Inertia

Critical inertia is the minimum level of system inertia that will ensure ERCOT's fast frequency responsive resources can be effectively deployed before frequency drops below 59.3 Hz *following the simultaneous loss of 2750 MW.*

Following a resource trip, the rate of the change of frequency (RoCoF) immediately after the event is solely a function of the inertia of the synchronous machines that are online and the magnitude of the generation loss. Load resources (LRs) with under-frequency relays providing RRS respond in about 0.5 seconds (or 30 cycles) after the frequency drops below their trigger level of 59.7 Hz.

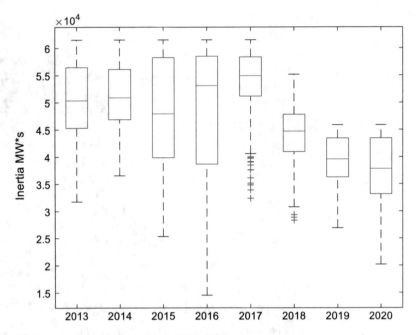

Fig. 7.7 Boxplot of inertia by coal units 2013–2020

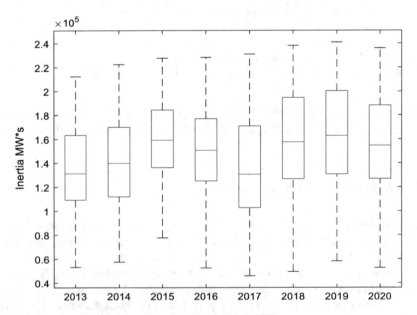

Fig. 7.8 Boxplot of inertia by combined cycle units 2013–2020

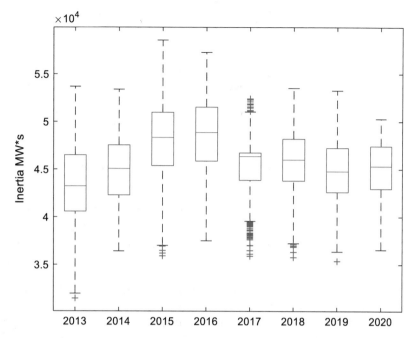

Fig. 7.9 Boxplot of inertia from PUNs 2013–2020

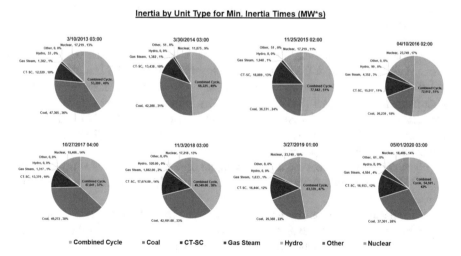

Fig. 7.10 Breakdown of inertia contribution by fuel type at the lowest inertia hours in 2013–2020

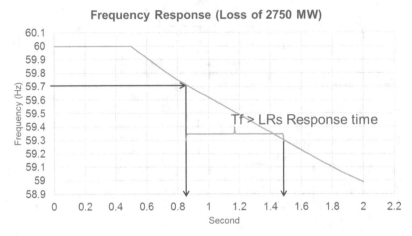

Fig. 7.11 Frequency response after loss of 2750 MW and time it takes for frequency to reach 59.3 Hz

ERCOT has defined critical inertia as the minimum inertia level at which a system can be reliably operated with current frequency control practices. Below critical inertia, frequency reserves from load resources may not have sufficient time to arrest system frequency before it reaches the first-stage UFLS trigger level (59.3 Hz in ERCOT). Thus, for ERCOT, critical inertia is the inertia level below which, for a generation loss of 2750 MW (the two largest units in the system), frequency takes less than 0.5 seconds to decline from 59.7 Hz to 59.3 Hz (i.e., there is insufficient time for load resources providing responsive reserve service (RRS) to respond and arrest frequency above UFLS trigger levels).

Figure 7.11 illustrates the concept by showing a frequency response after the loss of 2750 MW at low system inertia. In this case, it takes less than 0.5 seconds, denoted as T_f, for system frequency to drop from 59.7 Hz to 59.3 Hz.

ERCOT conducted a series of dynamic simulations for the cases with inertia conditions ranging from 108 GW·s to 376 GW·s to investigate how long it takes for frequency to drop from 59.7 Hz to 59.3 Hz when there is a 1150 MW of RRS from generation (no RRS from load resources are included). Figure 7.12 shows the corresponding frequency response.

Figure 7.13 summarizes the frequency travel time from 59.7 Hz (LR trip setting) to 59.3 Hz (UFLS setting) for each case if only 1150 MW of RRS from generation is in use.

A regression curve can be used to estimate the frequency travel time for the different inertia, which yields an approximation $T_f = 3 \cdot 10^{-5} \cdot \text{Inertia}^2 + 0.0016 \cdot \text{Inertia} + 0.0048$. Using the relationship, critical inertia was found to be 94 GW·s when T is 0.5 seconds, as shown in Fig. 7.14.

Based on this analysis and with a safety margin, ERCOT has identified its critical inertia to be 100 GW·s. Operating the grid reliably below this inertia level would require changes to ERCOT's current operating practices, which include, are but not

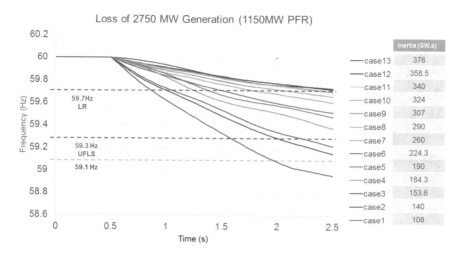

Fig. 7.12 Frequency response after loss of 2750 MW for 13 cases with different inertia levels (PFR-Primary Frequency Response)

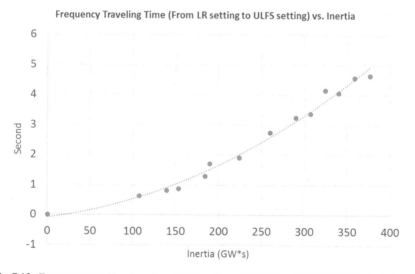

Fig. 7.13 Frequency travelling time from LR trip setting (59.7 Hz) to UFLS (59.3 Hz) for 13 inertia cases and trendline

limited to, faster FFR response time, higher FFR frequency trigger, and procuring inertia as an ancillary service. The lowest inertia experienced in ERCOT between 2013 and 2020 was 129 GW·s.

At inertia levels above critical inertia, ERCOT can ensure reliable operation by procuring a sufficient amount of RRS for expected inertia conditions.

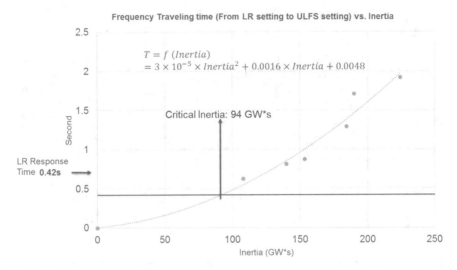

Fig. 7.14 Frequency travelling time from LR trip setting (59.7 Hz) to UFLS (59.3 Hz) at different inertia levels, trendline, and critical inertia

5 ERCOT Tools to Monitor and Forecast System Inertia

As continuous growth of wind and solar generation as well as coal retirements bring more uncertainties to grid operation, there is a great need to monitor system inertia in real time and also to help operators to predict it for the near future.

Monitoring of synchronous inertia and frequency deviation after the largest category C (N-2) event (2750 MW in ERCOT) were recommended measures from NERC Essential Reliability Services Working Group (ERSWG).[2] These measures have been passed on to NERC Resource Subcommittee in order to start data collection from all balancing authorities and interconnections in NERC.

In order to streamline monitoring and analysis of the system inertia as well as contribution by individual generation types, ERCOT developed a network model-driven inertia calculation in EMS by the end of 2016. Inertia calculation driven by the network model allows the automatic inclusion of new generation resources and removal of retired generation resources from the calculation. The calculation provides inertia contribution by unit type, PUN inertia, and total system inertia in real time with 4-second resolution. Total system inertia is monitored in real time in the

---

[2]The ERSWG had a multi-faceted purpose that includes developing a technical foundation of ERS; educating and informing industry, regulators, and the public about ERS; developing an approach for tracking and trending ERS; and formulating recommendations to ensure the complete suite of ERS are provided and available. The ERSWG reconciled a collection of analytical approaches and measures for understanding potential reliability impacts as a result of increasing variable resources and how those impacts can affect system configuration, composition, operation, and the need for increased ERS.

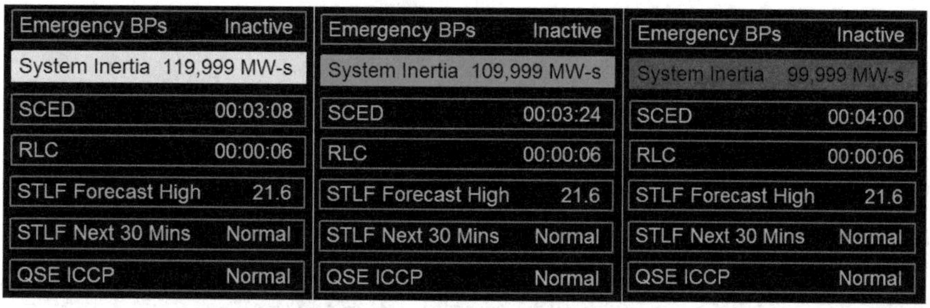

Emergency BPs	Inactive	Emergency BPs	Inactive	Emergency BPs	Inactive
System Inertia 119,999 MW-s		System Inertia 109,999 MW-s		System Inertia 99,999 MW-s	
SCED	00:03:08	SCED	00:03:24	SCED	00:04:00
RLC	00:00:06	RLC	00:00:06	RLC	00:00:06
STLF Forecast High	21.6	STLF Forecast High	21.6	STLF Forecast High	21.6
STLF Next 30 Mins	Normal	STLF Next 30 Mins	Normal	STLF Next 30 Mins	Normal
QSE ICCP	Normal	QSE ICCP	Normal	QSE ICCP	Normal

Fig. 7.15 Information an operator would see for different levels of low inertia in ERCOT

ERCOT control room, and a three-level alarming approach is implemented to ensure operator awareness. Figure 7.15 shows the information that an operator would see for different levels of low inertia. There is also an action plan in the control room for each of those levels:

- ≥110,000 MW·s to ≤119,999 MW·s—the monitor shows the value highlighted yellow.
- ≥100,001 MW·s to ≤109,999 MW·s—the monitor shows the value highlighted orange.
- ≤100,000 MW·s—the monitor shows the value highlighted red, and the operator has to take measures to restore system inertia to above 100,000 MW·s.

ERCOT operators will do the following:

- Monitor grid conditions closely when system inertia < 120 GW·s.
- Take actions when system inertia < 105 GW·s:

  - Target increasing system inertia > = 105 GW·s.
  - Possible actions:

    - Deploy Non-Spin from offline generation resources (including quick start generation resource (QSGRs) that carry Non-Spin). Resources that have historically carried offline Non-Spin can on an average provide ~4000 MW·s inertia increment.
    - Deploy remaining quick starts (not carrying Non-Spin). Quick start resources that have historically not carried Non-Spin can on an average provide ~6000 MW·s inertia increment.
    - Bring offline generation resources online that can be turned on within 1 hour by reliability unit commitment (RUC).

Additionally, a new RRS Sufficiency Monitoring tool was implemented in the ERCOT control room in March 2017. This tool conducts 24-hour look-ahead studies to assess the adequacy of procured RRS reserves for expected system inertia conditions derived using the latest current operating plans submitted by the

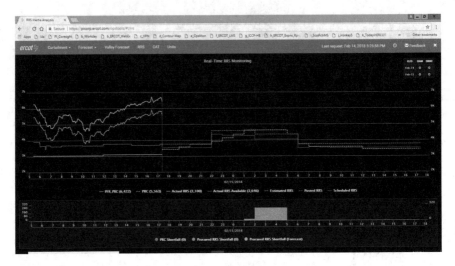

Fig. 7.16 Real-time RRS Sufficiency Monitoring

generation resources. This tool also has the ability to monitor sufficiency of available RRS reserves during real time based on actual system conditions. The Reliability Risk desk uses this tool and takes necessary actions per its current procedures. Though the tool was initially created for RRS Sufficiency Monitoring, underlying inertia forecast can be monitored with respect to 100 GW·s limit—see Fig. 7.16.

6 Impact of Parameter Changes on Critical Inertia

ERCOT's critical inertia is sensitive to parameters such as the response characteristics of the LRs that provide RRS, UFLS settings, and the size of the critical contingency. By changing one or more of these parameters, critical inertia can be set to a lower level. The following parameter changes were tested and are discussed below in detail:

- "Faster" frequency response for load resource providing RRS
- "Earlier" frequency response from load resources providing RRS
- Lower UFLS trigger
- Reduction of resource contingency

6.1 *"Faster" Frequency Response*

One way to reduce critical inertia would be to take advantage of the fast response of load resources providing RRS, which can provide full response in a few hundred milliseconds in the under-frequency events. Currently, full response from load resources is provided within 0.42 seconds after frequency reaches 59.7 Hz, but it is technically feasible for load resources to react more quickly. Several dynamic simulations were performed with load resources providing full response in 15 cycles rather than 25 cycles, while other parameters remain unchanged. In this scenario, critical inertia was reduced to 68 GW·s, as seen in Fig. 7.17.

6.2 *"Earlier" Frequency Response*

Another way to reduce critical inertia is to prepare system with frequency response resources that provide early response. ERCOT's load resources (LR) are currently activated at 59.7 Hz with a delay less than 50 ms. Under low inertia conditions, where frequency declines drastically for a loss of generation event, having earlier frequency response resources (higher-frequency set point), e.g., through a new fast frequency response (FFR) service, would slow down the frequency decline and allow more time for LR, providing RRS, to respond. As a result, critical inertia could be reduced.

One concern with FFR is a frequency overshoot. Since the triggering point is higher, the chance of such resources being deployed due to a relatively small event

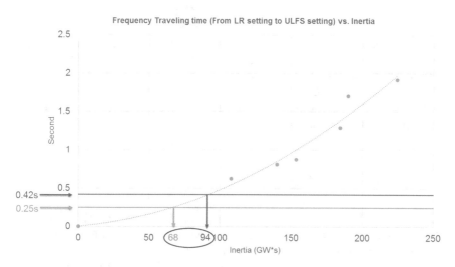

Fig. 7.17 Frequency travelling time vs. inertia and the impact on critical inertia of shortening LR response time

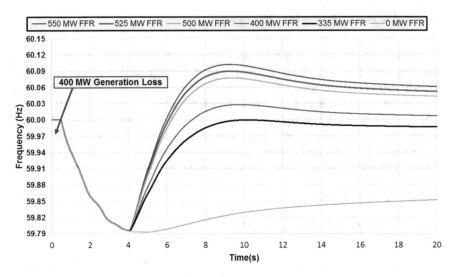

Fig. 7.18 Frequency overshoot with FFR

increases. For such small events, if MW of FFR deployed is much higher than the lost MW of generation, a frequency overshoot would be expected. Therefore, establishing a limit on FFR resources to avoid overshoot is necessary.

A low inertia condition (108 GW·s) power flow case is selected for dynamic simulations to determine the limit on FFR. Under a low inertia condition, it only takes a small generation loss to have FFRs triggered, and frequency overshoot is more severe. FFR is assumed to be triggered at 59.8 Hz with 25-cycle response time in this study. Simulation shows that under 108 GW·s inertia condition, the smallest generation loss that leads to frequency nadir at 59.8 Hz is 400 MW, shown as the "0 MW FFR" line in Fig. 7.18. For loss of 400 MW, a total of 335 MW FFR results in 0 overshoot above 60 Hz, whereas a total of 550 MW of FFR leads to an overshoot of 60.10 Hz (overshoot at or above 60.1 is in exceedance of NERC BAAL limit). The limit of 525 MW on FFR is established to keep overshoot under 60.09 Hz. With the help of 525 MW FFR to slow down frequency decay, system critical inertia is reduced to 88 GW·s as shown in Fig. 7.19. The impact on critical inertia of FFR with different response times is also studied and summarized in Fig. 7.20.

6.3   Lower UFLS Trigger

According to ERCOT Nodal Operating Guides 2.6.1, at least 25% of the ERCOT system load that is not equipped with high-set under-frequency relays shall be equipped at all times with provisions for automatic under-frequency load shedding. The under-frequency relays shall be set to provide load relief as follows (Tables 7.3 and 7.4):

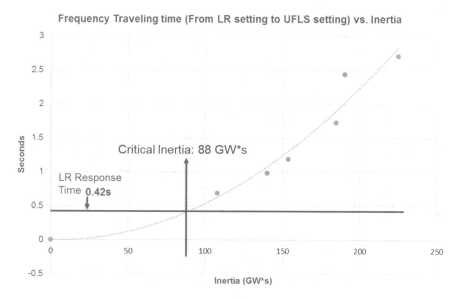

Fig. 7.19 Critical inertia with FFR

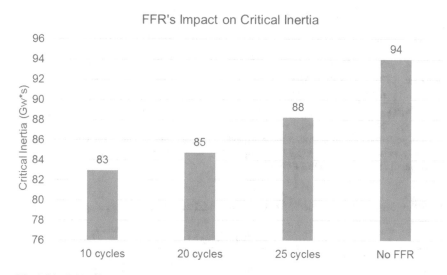

Fig. 7.20 Critical inertia with FFR (different response times)

If under-frequency relays are installed, these relays shall be set such that the automatic removal of individual generation resources from the ERCOT system meets the following requirements.

Table 7.3 ERCOT UFLS settings

Threshold	Load relief
59.3 Hz	5% of the ERCOT system load (total 5%)
58.9 Hz	An additional 10% of the ERCOT system load (total 15%)
58.5 Hz	An additional 10% of the ERCOT system load (total 25%)

Table 7.4 ERCOT generator under-frequency protection

Threshold	Load relief
Above 59.4 Hz	No automatic tripping (continuous operation)
Above 58.4 Hz up to and including 59.4 Hz	Not less than 9 minutes
Above 58.0 Hz up to and including 58.4 Hz	Not less than 30 seconds
Above 57.5 Hz up to and including 58.0 Hz	Not less than 2 seconds
57.5 Hz or below	No time delay required

Table 7.5 Comparison of UFLS settings for the systems experiencing low inertia issues

	Max (GW)	Min (GW)	Resource contingency criteria, under-frequency MW	UFLS frequency	UFLS frequency on 60 Hz reference
Nordic TSOs	70.00	25.00	1450	48.8	58.56
ERCOT	69.86	24.17	2752	59.3	59.3
NG UK	53.00	17.00	1000	49.2	59.04
Hydro-Québec	39.00	15.00	1700	58.5	58.5
ESKOM	35.00	19.00	930	49.2	59.04
Australia	30.40	14.30	760	49	58.8
EirGrid	6.40	2.30	500	48.85	58.62

Hydro-Québec's largest generation loss is set lower than 1700 MW during low inertia conditions
Source: ENTSO-E Report, "Future System Inertia 2"
ERCOT performed dynamic studies to investigate the impact from changing the UFLS first-stage triggering point from 59.3 Hz to 59.1 Hz. Figure 7.21 shows that this change allows for lowering critical inertia from 94 GW·s to 71 GW·s

Table 7.6 Impact on critical inertia from lowering UFLS trigger and/or shortening response times for load resources

	UFLS @59.3 Hz	UFLS @59.1 Hz
0.42 s LR response time	94 GW·s	71 GW·s
0.25 s LR response time	68 GW·s	52 GW·s

Table 7.5 reviews UFLS settings in different systems around the world experiencing low inertia issues. Since some of these systems operate at 50 Hz, UFLS settings are recalculated on a 60 Hz reference in the last column.

Table 7.6 summarizes the critical inertia impacts of lowering the UFLS trigger and/or shortening the response time for load resources providing RRS.

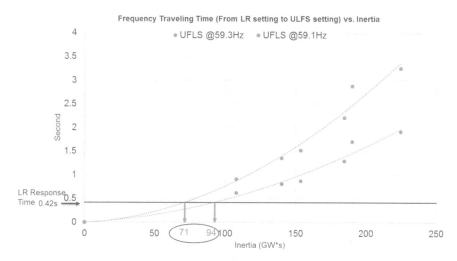

Fig. 7.21 Frequency travelling time vs. inertia and impact on critical inertia from lowering UFLS trigger from 59.3 Hz to 59.1 Hz

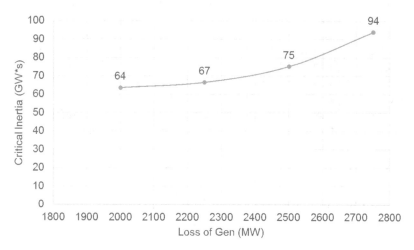

Fig. 7.22 Critical inertia at resource contingencies of different sizes from 2000 to 2750 MW

6.4   *Reduction of Largest Possible Loss of Generation*

The resource contingency criteria (RCC) for ERCOT, as defined in the NERC BAL-003 standard, is the simultaneous loss of the two nuclear units at the South Texas Project power plant, totaling 2750 MW.

ERCOT has conducted sensitivity studies to assess the impact of changing RCC on critical inertia. Figure 7.22 shows how critical inertia changes as a function of generation loss. It is recognized that further discussions will be needed to assess the feasibility of implementing this approach.

Low inertia conditions normally happen during shoulder seasons. It is possible that during these times some of the largest power plants are on planned maintenance for an extended period of time and, consequently, the largest possible resource contingency may be lower than 2750 MW. Thus, critical inertia can potentially be set at lower levels than 100 GW·s during those conditions as shown in Fig. 7.22.

7   International Review of Inertia-Related Challenges and Mitigation Measures

There are other systems in the world experiencing similar challenges associated with changes in their resource mix. Table 7.7 lists some relevant parameters of the systems currently facing low inertia issues. Note that Hydro-Québec and Nordic (Sweden, Norway, Finland, and part of Denmark) systems do not have particularly high capacity from non-synchronous generation (wind and solar); however, these systems are hydro generation based, which means inertia is relatively low (see Table 7.1) and primary frequency response is relatively slow. The size of the system's RCC with respect to the system size (i.e., minimum inertia level) is also an important factor, which indicates that South Australia system is at the forefront of this problem at the moment.

Table 7.8 lists different measures used by these systems to reduce critical inertia and/or keep inertia above critical levels. All of the systems monitor inertia and largest possible contingency in real time. Ireland and the United Kingdom additionally forecast inertia from the day ahead into real time, similar to ERCOT, as described in Sect. 5.

8   Summary of Potential Mitigation Measures to Lower Critical Inertia or Support Minimum Inertia Level

This section summarizes all possible technical solutions to mitigate low inertia issues. The advantages and disadvantages of the solutions are described as objectively as possible without favoring some solutions over others.

- *Faster and/or earlier response from load resources providing RRS*

    This solution allows for a reduced critical inertia level. Implementation requires coordinating with load resource owners to verify that faster response is possible. Changing the frequency trigger for LR response requires changing the settings of under-frequency relays, and it would also mean that LR will be deployed more often than currently and this may have adverse effects on LR participation in RRS.

Table 7.7 Countries experiencing low inertia challenges

System	Irelands	United Kingdom	Nordic	Quebec	South Australia
Peak demand, GW	6.4	53	70	39	3
Capacity from wind and solar	4 GW	>26 GW	10%	7%	35%
Minimum inertia, GW·s	20	135	125	60	2
Resource contingency criteria, MW	500	1000	1450	1700	650
Concerns	Lack of synchronizing torque, at RoCoF ≥0.5 Hz/s, significant amounts of non-synchronous generation will trip[a]	At RoCoF of 0.125 Hz/s, some non-synchronous generation will trip; at 1 Hz/s, all non-synchronous generation will trip	Slow PFR (hydro), time to UFLS is a concern	Low inertia (hydro), high RCC, slow PFR (hydro), time to UFLS is a concern	High (1–3 Hz/s) RoCoF after RCC, at which synchronous generation may trip and UFLS may malfunction

[a]Early on, the United Kingdom and Ireland enforced the use of loss of mains protection at non-synchronous generation plants in their distribution networks. The protection would trip wind or solar generation once a particular RoCoF was detected (0.125 Hz/s in the United Kingdom, 0.5 Hz/s in Ireland). This is because high RoCoF was used as an indication of an island forming in the distribution grid. As these systems saw more non-synchronous generation installed, system inertia declined and system-wide RoCoFs after large generation loss increased to a point where 0.125 Hz/s (or 0.5 Hz/s) RoCoF was no longer exclusively an indicator of an island in a distribution system. With RoCoF protection still in place at non-synchronous generation plants, this may potentially trip large amounts of non-synchronous generation further worsening the conditions on already perturbed system. The United Kingdom has increased the trigger on loss of mains protection from 0.125 Hz/s to 1 Hz/s for all resources installed after April 2014

- *New faster and/or earlier frequency response ancillary services product*
  Introducing a new separate from RRS fast frequent response product can potentially increase the pool of resources providing fast frequency response, including storage technologies.
- *Lower UFLS first step trigger*
  This solution allows for a reduced critical inertia level. This requires extensive testing and re-evaluation of UFLS schemes as well as ensuring that generators and demand customers are not adversely affected by potential operation at lower frequency levels for short periods of time.

Table 7.8 Mitigation measures to reduce critical inertia and to keep inertia above critical level

	Irelands	United Kingdom	Nordic	Quebec	South Australia	ERCOT
Monitor inertia and possible contingencies in real time	✓	✓	✓	✓	✓	✓
Forecasts inertia from DA into real time	✓	✓				✓
Dynamic assessment of reserves based on inertia conditions and largest resource contingency		✓				✓
Limit RCC based on inertia conditions	✓	✓		✓	✓	
Synchronous condensers (for inertia)	✓	✓			✓ (particularly looking at high inertia SCs)	
Enforce minimum inertia limit	✓	✓			✓ (for minimum inertia req.)	✓
Inertia market/auction/service inertia	✓				Y (for above minimum inertia levels)	
Faster-responding reserves	FFR	Enhanced frequency response service		Synthetic inertia from wind	"Contingency" FFR and "emergency" FFR	Y (LR providing RRS)

- *Bring more generation units online (RUC)*

  If system inertia is low, ERCOT may choose to bring more synchronous generation units online and thus increase system inertia. One benefit of this approach is that it would only be implemented during conditions when insufficient inertia was otherwise committed on the system, and therefore it would not affect market solutions during other periods. However, the generating units, once started, will run at least at their minimum sustainable level, producing energy that is not needed otherwise. This may result in renewable curtailment and adverse effects on energy prices. The generators committed for inertia would need to be uplifted (i.e., compensated) for their startup costs and power production at minimum generation level. It would be important to consider the relative inertia provided by the different units that could be brought online and their minimum generation levels in order to get the maximum inertia for the cost. This would be only recommended when the system inertia falls below critical inertia, which causes a serious concern for reliability.

- *Procure inertia as a separate ancillary service product*

  This mitigation measure would enable ERCOT to maintain necessary synchronous inertia levels through a direct market mechanism. Currently, inertia is provided by synchronous machines as the by-product of energy or ancillary service, and no compensation has been offered for this service. Once historic analysis of system inertia shows that ERCOT more frequently experiences extremely low inertia conditions, it could be more efficient to incentivize synchronous inertia service as an ancillary service. This requires changes to current market design. One possible synchronous inertial response (SIR) AS market construct was proposed during the ERCOT Ancillary Services Redesign project.

- *Install high inertia synchronous condensers*

  A synchronous condenser is similar to synchronous generator but without a turbine; its main purpose is to provide dynamic reactive power support. However, since it is a rotating machine, it can also contribute to total system synchronous inertia. This option can be implemented in a number of ways. If transmission service providers own the synchronous condensers, then their inertia contribution to total system inertia would need to be accounted for, but there would be no need for an inertia market.[3] If there is SIR market, some fossil-fueled generators may modify their generators to operate in synchronous condenser mode at night and sell inertia in those hours instead of energy and AS.

- *Synthetic inertial response requirement from wind generation*

  Another example of fast frequency response is synthetic inertia provision from Type 3 and 4 wind generation resources. In this construct, when a wind turbine plant controller senses a drop in system frequency, it extracts kinetic energy from the rotating mass of the wind turbine, which is seen from the grid as an increase in

---

[3] Recent studies for the Panhandle region proposed synchronous condensers for voltage support. As one of the long-term solutions, use of synchronous condensers for multiple purposes may prove to be more cost-efficient.

active power injection. The effectiveness of the response and recovery of wind generation resource to its pre-disturbance state depends on operating conditions of a wind generation resource. Therefore, this type of fast frequency response requires careful centralized coordination to ensure reliable system operation. While synthetic inertial response capability is already included as a part of interconnection requirement in Hydro-Québec, this technique has not been commercially utilized at a large scale. One option would be to combine synthetic inertial response requirements as a part of qualification requirements for the provision of RRS from wind generation, assuming there is interest from these resources to provide RRS in the future.

9   Conclusions

As more inverter-based resources are being connected to the power grid, the system inertia is declining. More load resources with under-frequency relays have been used to arrest the frequency excursions. However, as the system inertia becomes lower, the effectiveness of those load resources is deteriorated. Based on the operational experiences, ERCOT established a critical inertia, which is defined as the minimum level of system inertia that will ensure ERCOT's "fastest" load resources can provide sufficient frequency response before frequency hits 59.3 Hz at which the UFLS trigger. Based on the current set of frequency control practices, critical inertia in ERCOT is 100 GW·s.

ERCOT continues to explore more solutions to mitigate the low inertia issue, by developing decision support tools to monitor inertia conditions in real time, creating fast frequency response as ancillary service product, and allowing wind and solar resources to provide frequency responsive capability.

References

1. ENSTO-E Report, "Future System Inertia", project report by Nordic Transmission System Operators
2. Eto, Joseph H. "Use of frequency response metrics to assess the planning and operating requirements for reliable integration of variable renewable generation." *Lawrence Berkeley National Laboratory* (2011).
3. Essential Reliability Services Task Force (ERSTF) and Essential Reliability Services Working Group, NERC, 2014-2016.
4. Prabha Kundur, Power Systems Stability and Control, McGraw Hill, 1993

Chapter 8
Multiple-Period Reactive Power Coordination for Renewable Integration

Nomenclature

AVC	Automatic voltage control
COP	Current operating plan
EMS	Energy management system
ERCOT	Electric Reliability Council of Texas
GE	General Electric
ISO	Independent system operators
LP	Linear programing
MILP	Mixed-integer linear programming
NERC	North American Electric Reliability Corporation
QSE	Qualified scheduling entity
RTCA	Real-time contingency analysis
RTNET	Real-time network analysis (state estimator)
SCADA	Supervisory control and data acquisition
SCOPF	Security-constrained optimal power flow
SE	State estimator
SMTNET	Multi-time point study network analysis
STNET	Study network analysis
TO	Transmission operator
TSP	Transmission service provider

Symbols

n	Index of controllable reactive devices
t	Index of time
m	Index of base and contingency case
k	Index of buses monitored for voltage limit
T	Number of time steps

N	Number of controllable reactive devices
M	Number of base and contingency case
K	Number of buses monitored for voltage limit
I	Set of buses
K_{c}	Set of controllable capacitors
K_{r}	Set of controllable reactors
K	Set of buses monitored for voltage limits
$C_{n,t}$	Status of the n-th controllable reactive devices at time interval t
$W_{n,t}$	Cost of changes made to n-th controllable reactive devices at time interval t
$P^{\mathrm{G}}_{i,t,m}/P^{\mathrm{L}}_{i,t,m}$	Generation/load active power at the i-th bus for the m-th case at time interval t
$Q^{\mathrm{G}}_{i,t,m}/Q^{\mathrm{L}}_{i,t,m}$	Generation/load reactive power at the i-th bus for the m-th case at time interval t
$G_{ij,t,m}/B_{ij,t,m}$	Conductance/susceptance of line connecting bus i and j for the m-th case at time interval t
$\delta^{\mathrm{Hi}}_{k,m,t}/\delta^{\mathrm{Low}}_{k,m,t}$	Relax of voltage violation constraints of k-th bus for the m-th case at time interval t
$V^{\mathrm{Hi}}_{k,m,t}/V^{\mathrm{Low}}_{k,m,t}$	High/low voltage limit of k-th bus for the m-th case at time interval t
$S^{k}_{n,m,t}$	Sensitivity of n-th controllable reactive devices to k-th bus voltage magnitude for the m-th case at time interval t
$\pi_{k,m,t}$	Penalty of relax of voltage limit constraints
ρ_n	Temporal constraint of n-th controllable reactive devices
$\gamma_{\mathrm{max},n}$	Maximum number of operations allowed for n-th controllable reactive devices

1 Introduction

The Electric Reliability Council of Texas, Inc. (ERCOT) is an independent system operator certified by the Public Utility Commission of Texas pursuant to Section 39.151 of the Texas Public Utility Regulatory Act and is responsible for ensuring the reliability and adequacy of the regional electrical network. The ERCOT region is approximately 200,000 square miles (75 percent of the land area in Texas), with a record peak demand at 74,820 MW. The ERCOT region has approximately seven million retail customers [1]. To support its operations, ERCOT has multiple systems which include energy management system (EMS) and market management system (MMS) [2]. In EMS, there are several application functions running both in real time and in study mode. The real-time applications include state estimator (also referred to as real-time network analysis—RTNET), contingency analysis (CA), and dynamic stability analysis tools. Together, these tools detect any violation of the operational constraints and develop the mitigation plans when needed. The deployment of these tools has also become an industry standard practice to ensure the power system reliability and security.

As the voltage plays a very important role in power delivery and reliability, the voltage magnitude of the power grid is assessed under both normal and contingency conditions and monitored by the ERCOT operators in real time [2, 3]. In this regard, the control of reactive power devices is essential to keep the voltage in the base case and the post-contingency states within a desired range.

The voltage control was coordinated between ERCOT and transmission operators (TOs). ERCOT and TOs conduct seasonal studies to determine a seasonal voltage profile for the generation resources to control their terminal voltage. Some TOs alter the voltage set point in real-time operations from the seasonal voltage profile. This is done to manage varying voltage conditions that could occur throughout the day as the switching of shunt capacitors alone is insufficient to change the varying reactive power flows. Operators at ERCOT and TOs rely on the past experiences to heuristically determine when and how to switch the shunt reactive power devices (both capacitors and reactors) and change the tap position of the transformers. However, this practice is inefficient when the penetration of renewable resources becomes high [2, 3].

A few challenges need to be overcome to ensure the power grid voltage security. First, operators should consider the temporal constraints and the limits on the number of operations when switching on/off these reactive power devices. Second, the schedule of reactive power resources needs to be coordinated between continuous and discrete devices. Third, the study should be performed for multiple time intervals to provide a look-ahead capability to assess the reactive power control requirements for the future hours.

2   Literature Review

Extensive works have been done on the subject of the reactive power control and scheduling for power systems. The real and reactive power can be decoupled in the neighborhood of the solution to accelerate the solution, and this property can be exploited to form a linear relationship between the controls and constraints [4]. The mathematical models of power system components are also critical in such an analysis, and more details can be found in [5]. To void divergence due to ill-condition at the steady-state voltage stability limit (critical point) of the system, a method of finding a continuum of power flow was proposed in [6].

Recently, the control of the reactive power has been formulated as a security-constrained optimal power flow (SCOPF) problem, which is a nonlinear and nonconvex optimization problem [7–9]. With the size of the system and the number of constraints growing, the problem becomes even more challenging. The work in [7] provides an overview of optimal reactive power planning and an exhausted list of methods that have been used or suggested to solve the SCOPF problem. The reference [8] presents a survey on the strengths and weaknesses of different models proposed previously to solve optimal reactive power dispatch. The emerging challenges and research opportunities for voltage control in smart grids were also reviewed in [9], from both transmission and distribution grid perspective.

SCOPF improves the voltage regulation in real time, but the objectives can differ in its implementations. The reliability and efficiency can be increased by coordinating the operation of reactive power resources. This is achieved by solving a centralized optimization problem. A two-level voltage controller for large-scale power systems was proposed in [10] with an objective of maintaining a near-optimal voltage profile by coordinating discrete reactive power control devices [10]. The authors in [11] determined the minimum amount of shunt reactive power support which indirectly maximizes the real power transfer before voltage collapse is encountered. A voltage stability index is used as an indirect measure to the closeness of reaching the steady-state voltage stability limits. The work in [12] developed a control method to determine the most effective control actions in order to re-establish critical reactive power reserves across the system while maintaining a minimum amount of voltage stability margin. An optimized reactive reserve management scheme based on the optimal power flow was proposed in [13] and solved by the Benders decomposition methodology. The authors described a set of algorithms for SCOPF and their configuration in an integrated package for real-time security enhancement [14]. The SCOPF solutions incorporate base case and contingent state load flow solutions, network switching, staged constraint and control activation, and linear program solutions. A two-level decomposition technique for Var sources planning in electric power systems was presented in [15], in which total flexibility in the corrective or preventive treatment for control variables is provided.

In Europe and China, the progresses were made in the development and deployment of closed-loop, system-wide automatic voltage control (AVC) system. The hierarchical control consists of primary voltage control, secondary voltage control, and tertiary voltage control. A new AVC technology was proposed to improve the reliability and robustness of coordinated voltage regulation in the transmission networks [16]. An AVC scheme based on adaptive zone division was introduced in the chapter [17]. The control zones are reconfigured online and updated in accordance with variations in the grid structure. The secondary voltage control in North America is still a manual process except in ISO New England [18] and PJM [19]. In [18], the authors studied the minimal information structure needed for monitoring and control of the voltage profile and the reactive power flow of a power system. The optimal voltage control system was developed to greatly improve both the pre- and post-contingency performance in PJM [19].

As renewable resources are replacing fossil fuel in the power system, the voltage could fluctuate more quickly and frequently. To prevent the wind units from cascading trip faults, the work in [20] discussed a robust master-slave two-level coordinated static voltage security region (VSR), whereas the master system-wide VSR aims to coordinate centralized wind farms to ensure security under both normal operating conditions and $N - 1$ contingency conditions. A hierarchical voltage control methodology was discussed in [21] to optimize the reactive reserve of a power system with high levels of wind penetration to increase the capability to manage a future AC/DC hybrid network. In [22], the dynamic reactive power optimization was proposed to minimize the total transmission loss by coordinating the continuous and discrete reactive power compensators while guaranteeing the specific physical and operating constraints.

Recent advancement in machine learning also spurred interest in applications of machine learning techniques to solve optimal reactive dispatch problems. In [23], setting the tap positions of load tap changers (LTCs) for voltage regulation in radial power distribution systems under uncertain load dynamics was formulated as a Markov decision process and solved by a batch reinforcement learning algorithm. The work in [24] built a data-driven, model-free, and closed-loop control, trained using deep reinforcement learning algorithms to derive autonomous voltage control strategies.

However, all of previous works solved a SCOPF problem only for a single time interval except in [22, 25, 26]. The dynamic reactive power optimization was formulated for multiple-period, but it was mainly applied to the distribution network. The work in [25] investigated a look-ahead optimal reactive power dispatch. The work presented in [26] optimized the reactive power dispatch schedule for a future look-up horizon, and the voltage security-constrained multi-period optimal reactive power flow was solved using benders and optimality condition decompositions. However, both the works in [25, 26] attempt to solve a computation-intensive problem.

Despite the works reported in [25, 26], the voltage control problem for a bulk power system remains a great challenge since it deals with a set of both continuous and discrete controls. The problem becomes even more complex and larger in size when we expand it to a multiple-hour reactive power scheduling problem. This chapter describes a new framework to solve a multiple-period reactive power schedule problem which is computationally tractable and can be solved by a commercial optimization solver.

3 Renewable Integration and New Transmission Operators

The last decade has seen a dramatic increase in the penetration of renewable generation resources in ERCOT. The wind installed capacity has grown from ~10 GW in 2011 to more than 30 GW by the end of 2020. Likewise, as of 2016, solar generation is at 556 MW and projected to increase to about 9 GW in 2022. Due to the zero fuel cost, renewable generation will typically be most economical and will displace other thermal generation resources from dispatch. This causes a fundamental shift in the type of generation dispatched online and associated power flows. The real power to serve load can be transferred further away with more economical sources, but since reactive power cannot efficiently travel in a long distance, they must be supplied locally or near the reactive loads. Due to the variability of renewables, the reactive power flows can change quickly, thus making it more challenging to manage and schedule the reactive power resources [3].

The integration of renewables and associated transmission upgrades has progressively introduced several new TOs with the ERCOT regions over the last 10 years. These TOs manage voltage in real time but do not perform future look-ahead analysis or real-time post-contingency analysis due to limited access to the model and data. ERCOT is solely responsible for those reliability tasks to meet North American Electric Reliability Corporation (NERC) reliability standard requirements. TOs will implement corrective actions either by ERCOT direction or independently after coordination or consultation with ERCOT.

4 Reactive Power Coordination (RPC) Tool

To address the aforementioned challenges associated with reactive power control and facilitate the coordinative schedule among TOs, a multiple-hour reactive power coordination (RPC) is the most efficient solution to solve voltage and reactive power flow violations due to three reasons:

1. It minimizes the number of real-time voltage and voltage stability limit violations (i.e., enhanced reliability).
2. It helps to ensure that sufficient reactive reserves are maintained.
3. It minimizes the number of control movements.

The RPC tool will solve a multiple-hour optimization problem and will aid ERCOT in compliance to NERC Reliability Standards for Operational Planning Analysis. A detailed discussion on the architectures, objective functions, constraints, and solution methodology of RPC is presented as follows.

4.1 Architecture of RPC

SMTNET and STNET applications from the General Electric (GE) EMS 3.0 applications suite are selected for the implementation of RPC due to their capabilities in handling both continuous and discrete control variables and a large number of contingencies. SMTNET is a study tool where the case preparation for the look-ahead study is performed. SMTNET and STNET are also a part of EMS functions which ERCOT uses on a daily basis to assess the grid security and to perform the contingency analysis.

The overall procedure of multiple-hour RPC is depicted in Fig. 8.1a, which is composed of two components: the case preparation and the execution of RPC.

In the first step, the case preparation involves the following four procedures, as shown in Fig. 8.1a:

- Load network model.
- Read load forecast (LF).

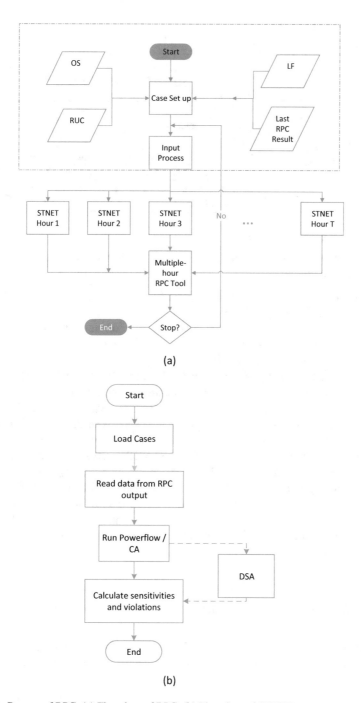

Fig. 8.1 Process of RPC. (**a**) Flowchart of RPC. (**b**) Flowchart of STNET

Table 8.1 Input data to RPC

Category	Data
Future data	Load forecasts
	Reliability unit commitment (RUC)/unit COP data
	Outages
	DC tie schedules
	Wind/solar generation forecast
Model data	Voltage limits
	Acceptable number of switches for shunts/time period (temporal)
	Reactive capability—shunts, generator capability curves, etc.
	Voltage schedules (target voltage and tolerance band)
	Load distribution factors

- Read outage schedule data.
- Read the current operating plan (COP)[1] and reliability unit commitment (RUC) data for the time of study.

The data input to RPC is described in details in Table 8.1, which includes both the model data and future operational data.

Once the first step is completed, not only the current state of the system is determined, but also the future and model data will be loaded to build power flow cases for each hour of the future time of study.

In the second step, the proposed framework creates the T STNET cases and solves the corresponding power flow. The more details can be found in [2]. Moreover, it also calculates the sensitives of the control variables with respect to the voltage violation constraints. Both the solved power flow and the sensitives are then used as the input to the multiple-hour RPC to minimize the objective function while respecting all of the operational constraints. These STNET processes are run in parallel to improve the performance. Each STNET process runs a contingency analysis and writes out the constraints for each time point and the controls with sensitivities; see Fig. 8.1b. With the results from each STNET run, the multiple-hour RPC will be executed. The program will stop if the termination criterion is satisfied. Otherwise, the power flow and the sensitivities will be resolved at the updated state, and the process iterates.

4.2   Objective Function of RPC

The primary objective function of this RPC tool consists of multiple components, which include (1) to minimize the cost of changes made to controllable reactive

---

[1]COP reflects "expected" operating conditions of generation units for the future hours.

devices, (2) to minimize the number of shunt switching control actions, and (3) to reduce the number of voltage violations in the network.

The controllable variables used to minimize the aforementioned objective functions are the voltage set points of the generation units, the switching status of shunt capacitors and reactors, and the tap positions of the load tap changers' transformers. It should be noted that other objective functions and controllable variables can be added if needed due to the flexibility in the proposed framework.

4.3   Constraints of RPC

The RPC tool incorporates the following constraints in the optimization problem.

1. System Voltage Limits

   The system voltage limits must be maintained within their normal (pre-contingency) and emergency (post-contingency) limits. The default limits are set from 0.95 p.u. to 1.05 p.u. for normal conditions and 0.9 p.u. to 1.1 p.u. for emergency conditions. Some system voltage limits are different as provided by the TO or facility owner.

2. Reactive Reserves

   Dynamic reactive reserves must be maintained above ERCOT-defined minimum reactive reserve limits. These limits will vary from one reactive zone/area to another since some buses and reactive controls have a greater affect than others. These minimum reactive reserve limits will be determined through separate studies and analysis but will be no lower than 5% of the reactive power loading (customer demand plus losses) in the reactive zone or the single largest reactive power producing source. Single station reactive power loading should be no greater than 80% in the pre-contingency case.

3. Physical Constraints

   The reactive capability of controls must be utilized to provide realistic response to contingencies (e.g., reactive capability curves for generation resources, reactive capability of static Var compensator (SVC), automatic voltage regulator (AVR) status, etc.).

4. Temporal Constraints for Switching Devices

   Temporal constraints must be utilized to provide feasible operational actions for all reactive controls. These constraints are the minimum startup time, the minimum runtime (also lead time), and the maximum number of operations within 24 hours.

4.4   Mathematic Formulations of RPC

The mathematic formulation of the proposed RPC problem is as follows. The objective function is

$$\text{Min} \sum_{t=1}^{T} \sum_{n=1}^{N} (C_{n,t} \cdot W_{n,t}) + \sum_{t=1}^{T} \sum_{n=1}^{N} |C_{n,t} - C_{n,t-1}|$$
$$+ \sum_{t=1}^{T} \sum_{k=1}^{K} \sum_{m=0}^{M} \left(\delta_{k,m,t}^{Hi} \cdot \pi_{k,m,t} \right) \tag{8.1}$$
$$+ \sum_{t=1}^{T} \sum_{k=1}^{K} \sum_{m=0}^{M} \left(\delta_{k,m,t}^{Low} \cdot \pi_{k,m,t} \right)$$

subject to:

1. Power flow constraints

$$P_{i,t,m}^{G} - P_{i,t,m}^{L} = V_{i,t,m} \sum_{j=1}^{J} V_{j,t,m} (G_{ij,t,m} \cos \theta_{ij,t,m} + B_{ij,t,m} \sin \theta_{ij,t,m}) \tag{8.2a}$$
$$m = 0 \cdots M \quad t = 1 \cdots T \quad i \in I$$

$$Q_{i,t,m}^{G} - Q_{i,t,m}^{L} = V_{i,t,m} \sum_{j=1}^{J} V_{j,t,m} (G_{ij,t,m} \sin \theta_{ij,t,m} - B_{ij,t,m} \cos \theta_{ij,tm}) \tag{8.2b}$$
$$m = 0 \cdots M \quad t = 1 \cdots T \quad i \in I$$

where $m = 0$ represents the base case and $m = 1 \cdots M$ represents the contingency cases.

2. Bus voltage constraints for the monitored bus $k \in K$ at the interval t

$$\sum_{n=1}^{N} S_{n,m,t}^{k} (C_{n,t} - C_{n,t}^{0}) \leq V_{k,m,t}^{Hi} + \delta_{k,m,t}^{Hi} \tag{8.3a}$$

$$\delta_{k,m,t}^{Hi} \geq 0 \tag{8.3b}$$

$$\sum_{n=1}^{N} S_{n,m,t}^{k} (C_{n,t} - C_{n,t}^{0}) \geq V_{k,m,t}^{Low} - \delta_{k,m,t}^{Low} \tag{8.3c}$$

$$\delta_{k,m,t}^{Low} \geq 0 \tag{8.3d}$$

$$k \in K \quad m = 0 \cdots M \quad t = 1 \cdots T$$

3. Temporal constraint for the maximum number of switching allowed in the entire study period

$$\sum_{t=1}^{T} |C_{n,t} - C_{n,t-1}| \leq \gamma_{\max,n} \quad n = 1 \cdots N \tag{8.4}$$

4. Temporal constraints of reactive devices

$$\rho_{n,t}(C_{n,1}, C_{n,2}, \cdots C_{n,T}) = 0 \quad n = 1 \cdots N \quad t = 1 \cdots T \tag{8.5}$$

5. Reactive reserves

$$\sum_{i \in G} \left(Q_{i,t,0}^{G} - Q_{i,t,0}^{G} \right) \geq Q_t \tag{8.6}$$

4.5   Solution Methodology

The formulated RPC problem can be generally described in (8.1)–(8.6), where a more detailed discussion on (8.5) will be presented in Sect. 7. Since the controllable variables involve both continuous and discrete control actions, the formulated RPC is a nonlinear mixed-integer optimization problem. Various methods proposed to solve a non-convex nonlinear mixed-integer optimization problem have been compared in [7, 8]. The combinatorial methods can produce discrete solutions which are superior to others in terms of reducing the objective function. However, their computing time is prohibitive for real-time execution.

To meet the real-time requirement of coordinating the reactive power schedule for a grid like ERCOT, the method to solve the RPC needs to be fast and robust in its capability in handling different scenarios. To this end, linear programming (*LP*) is adopted so that the RPC problem can be solved by a commercial optimization solver, e.g., CPLEX. The discrete control actions are modeled as continuous control actions within the *LP* framework, which is followed by a discretization after the continuous solution is found. If the number of controls requiring discretization is high, a more heuristic method is used. More details for the solution methodology can be found in [14].

5   Special Considerations

5.1   Sensitivity of Reactive Power for a Regulating Bus

The reactive power limits of the generation units can be modeled as a *D*-shaped curve [5]. These are used in the power flow and contingency analysis. If a unit reaches its reactive power limit, the bus to which the unit is connected will be switched from a *PV* bus to a *PQ* bus. However, this will make the system less stable and may cause some convergence issues when solving a power flow (the cases will be unsolved). This problem can be overcome by using the limits in the *LP* problem. However, since the reactive power for a regulating bus is unknown, we cannot compute the sensitivities relative to the activated constraints in the *LP*. A remedy to this is to choose the line or transformer that is connected in series with the corresponding unit and then model the reactive power flow (Q) across the line as a constraint and set the flow limits as the unit reactive power limits.

5.2 Reactive Device's Temporal Constraint

A new method of modeling the temporal constraints for the shunt switching devices is proposed here, which is detailed as follows.

Case 1: Initial state of a capacitor is *OFF*

The temporal constraint at the time interval t for a minimum downtime t_m is given by

$$C_{i,t} - C_{i,t-1} + C_{i,t-t_m} \geq 0 \quad t = 1 \cdots T \tag{8.7a}$$

$$C_{i,t} - C_{i,t-1} + C_{i,t-t_m} \leq 1 \quad t = 1 \cdots T \tag{8.7b}$$

Case 2: Initial state of a capacitor is *ON*

The temporal constraint at the time interval t for a minimum uptime t_m is given by

$$C_{i,t} - C_{i,t-1} + C_{i,t-t_m} \leq 0 \quad t = 1 \cdots T \tag{8.8a}$$

$$C_{i,t} - C_{i,t-1} + C_{i,t-t_m} \geq -1 \quad t = 1 \cdots T \tag{(8.8b)}$$

Case 3: Initial state of a reactor is *ON*

The temporal constraint at the time interval t for a minimum uptime t_m is given by

$$C_{i,t} - C_{i,t-1} + C_{i,t-t_m} \geq 0 \quad t = 1 \cdots T \tag{8.9a}$$

$$C_{i,t} - C_{i,t-1} + C_{i,t-t_m} \leq 1 \quad t = 1 \cdots T \tag{(8.9b)}$$

Case 4: Initial state of a reactor is *OFF*

The temporal constraint at the time interval t for a minimum downtime t_m is given by

$$C_{i,t} - C_{i,t-1} + C_{i,t-t_m} \leq 0 \quad t = 1 \cdots T \tag{8.10a}$$

$$C_{i,t} - C_{i,t-1} + C_{i,t-t_m} \geq -1 \quad t = 1 \cdots T \tag{8.10b}$$

5.3 Handling Special Capacitor Banks

In traditional implementations, the shunt devices are modeled as the power injections at a bus, and the breakers are not used. In some cases, the switching of the shunt

Fig. 8.2 Special
capacitor bank

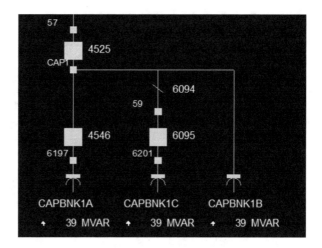

devices at a bus has to be in a particular sequence. Consider the configuration shown
in Fig. 8.2 as an example. The capacitor CAPBNK1 is controlled by the switch 4525
so that CAPBNK1B has to be connected to the grid before CAPBNK1A and
CAPBNK1C can be switched in. In the traditional implementations, this has not
been modeled. In this section, we create two additional constraints to dictate the
order of switching along with the other operational constraints.

Assume that the status of the master and slave capacitor shunt device is given by
$C_{m,t}$ and $C_{s,t}$, respectively. When the master capacitor is switched on, the switching
order constraint is given by

$$C_{m,t} - C_{s,t} \geq 0 \qquad\qquad (8.11a)$$

When the master capacitor is switched off, the switching order constraint is given
by

$$C_{m,t} - C_{s,t} \geq -1 \qquad\qquad (8.11b)$$

6 Case Studies

6.1 ERCOT Network Model

Every 5 minutes, ERCOT is running a state estimation to solve a power flow case,
which is sequentially used for contingency analysis. The ERCOT network model
consists of over 700 generation units and 8000+ buses. The contingency list contains
7976 contingencies. In this section, the effectiveness of the proposed method is
validated by using ERCOT model and data.

Table 8.2 States for the ten capacitors between T_1 and T_{10}

$C_{n,t}$	T_1	T_2	T_3	T_4	T_5	T_6	T_7	T_8	T_9	T_{10}
#1	0	0	0	0	0	0	0	0	0	0
#2	0	1	1	0	1	0	0	0	0	0
#3	0	1	0	0	0	0	0	0	0	0
#4	0	0	0	0	0	0	0	0	0	0
#5	0	0	0	0	1	1	1	1	1	0
#6	0	0	0	0	1	1	0	0	1	1
#7	0	0	0	0	0	0	0	0	0	0
#8	0	0	0	0	1	1	1	1	0	0
#9	0	0	0	0	0	0	1	1	0	0
#10	0	0	0	0	0	0	0	0	0	0

6.2   Verification of Temporal Constraints

Before (8.7)–(8.11) were implemented, they were tested to ensure that they can capture all improper state transitions. This will ensure that the constraint defined in (8.7)–(8.11) will add a penalty to the objective function and enforce proper state transitions.

6.2.1   Initial State of Capacitor Is OFF

Assume that there are ten capacitors (denoted as "#1–#10") with an initial state "OFF".[2] Without a loss of generality, both the minimum up- and downtime for these capacitors are assumed to be 2 hours. One example of the state for these ten capacitors within the next 10 hours is provided in Table 8.2. Using the states in Table 8.2, the temporal constraints against the state transitions can be calculated accordingly as shown in Table 8.3.

It is noted that a value of -1 in Table 8.3 represents a violation of the minimum uptime requirement. We can see a violation in the hour T_6 for the capacitor "#2." The reason is that the capacitor "#2" is switched back to online status in the hour T_5 and then switched off in the hour T_6. Thus, this operation does not meet the minimum uptime requirement. Another example is seen for the capacitor "#3" in the hour T_3. Similarly, there is a violated state transition for the minimum downtime in the hour T_5 for the capacitor "#2." The transition from the hour T_4 to T_5 does not respect the minimum downtime so that the state constraint for that hour is 2. This represents a failure of the constraint (8.7). If we incorporate this constraint in *LP*, then a penalty is enforced on the objective function, and CPLEX can thus find a solution without violating this transition constraint.

---

[2]For a capacitor, 1 indicates that the capacitor is on and 0 indicates a capacitor is switched OFF.

Table 8.3 State constraints of the ten capacitors

	$C_2 - C_1$	$C_3 - C_2 + C_1$	$C_4 - C_3 + C_2$	$C_5 - C_4 + C_3$	$C_6 - C_5 + C_4$	$C_7 - C_6 + C_5$	$C_8 - C_7 + C_6$	$C_9 - C_8 + C_7$	$C_{10} - C_9 + C_8$
#1	0	0	0	0	0	0	0	0	0
#2	1	0	0	2	-1	1	0	0	0
#3	1	-1	1	0	0	0	0	0	0
#4	0	0	0	0	0	0	0	0	0
#5	0	0	0	1	0	1	1	1	0
#6	0	0	0	1	0	0	1	1	0
#7	0	0	0	0	0	0	0	0	0
#8	0	0	0	1	0	1	1	0	1
#9	0	0	0	0	0	1	0	0	1
#10	0	0	0	0	0	0	0	0	0

Table 8.4 Switching sequence of master and slave capacitors

	Master	Slave1	Slave2		
	C_1	C_2	C_3	$C_1 - C_2 \geq 0$	$C_1 - C_3 \geq 0$
Master OFF	0	0	0	0	0
Master ON	1	0	0	1	1
Master ON	1	0	1	1	0
Master ON	1	1	0	0	1
Master ON	1	1	1	0	0

6.2.2   Sequence Constraints Between Master and Slave Capacitors

The logic implemented in (8.7–8.11) was also validated by another example. The initial statuses for one master capacitor and two downstream slave capacitors are offline. The master capacitor can be switched to be online with four possibilities for the statues for two slave capacitors. The sequence constraint is observed in this case as the constraint is satisfied when the master capacitor is switched online first—see Table 8.4.

6.3   Simulation Results

Two ERCOT cases are used to demonstrate the effectiveness of the proposed RPC tool. The first test case is a summer case with a modest load and a low wind generation. Figure 8.3 shows the load and online generation capacity for this operational day.

Twelve base cases between 7:00 a.m. and 6:00 p.m. were created to test RPC. The forecasted load and the scheduled generation capacity are given in Table 8.5. The actual generation dispatch and load are also provided for a reference.

The optimization horizon is configured to be 12 hours. Before the execution of optimization, a large number of violations for both base cases and contingencies already exist, especially during the first morning hours when the load ramps up. Table 8.6 gives the number of voltage violations for the 12 time snapshots in the study before and after optimizing reactive power schedule. The number of violations for all of 12 hours has been significantly cut down after being solved by RPC. The effectiveness of the proposed method is also validated for the contingencies. The only exceptions are the hour of 11:00 and 12:00. This may be caused by a different dispatch pattern which is hard to precisely predict in the morning. The same conditions are also used to run single-hour RPC tool for 12 hours starting at 7:00 a.m. Table 8.7 shows a comparison between those cases solved with the simple-hour optimization and multiple-hour optimization. Multiple-hour RPC is more advantageous as it reduces the number of controls moved to 128 compared to the

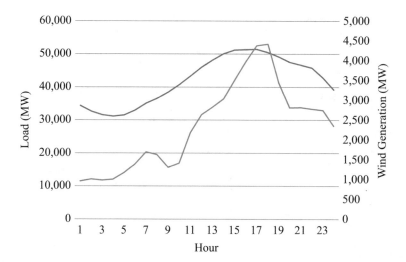

Fig. 8.3 Load and wind generation profile on June 5, 2017 (red, load; blue, wind generation)

Table 8.5 Test case for June 5, 2017

Time point	Online generation (MW)	Load forecast (MW)	Actual generation (MW)	Actual load (MW)
6/5/2017 7:00	36,693	36,517	35,665	34,993
6/5/2017 8:00	38,716	38,601	37,624	36,893
6/5/2017 9:00	41,212	40,487	40,051	39,296
6/5/2017 10:00	44,181	43,238	43,029	42,199
6/5/2017 11:00	47,160	45,405	46,045	45,103
6/5/2017 12:00	49,725	47,628	48,869	47,650
6/5/2017 13:00	52,041	50,456	50,964	49,887
6/5/2017 14:00	53,519	51,784	52,448	51,310
6/5/2017 15:00	54,089	52,694	52,998	51,874
6/5/2017 16:00	53,978	51,990	52,929	51,778
6/5/2017 17:00	53,434	50,392	52,391	51,274
6/5/2017 18:00	51,977	49,801	50,892	49,867

multiple single-hour RPC runs. Especially, single-hour RPC does not consider the temporal constraints of switching shunts. As a result, it may cause the same shunt device to switch more frequently. In terms of unsolved constraints, both multiple-hour RPC and single-hour RPC yield a similar performance.

The second test case is a low-load and high-wind case, on March 17, 2018. The load and wind generation profiles are shown in Fig. 8.4. The wind generation ramped up from 4149 MW at 16:00 to 8072 MW at 21:00.

Table 8.6 Voltage violations for the cases on June 5, 2017

Time point	Base case violations before RPC	CTG violations before RPC	Base case violations after RPC	CTG violations after RPC
6/5/2017 7:00	464	51	1	0
6/5/2017 8:00	395	27	8	1
6/5/2017 9:00	345	24	7	0
6/5/2017 10:00	136	6	3	0
6/5/2017 11:00	114	1	10	3
6/5/2017 12:00	64	0	6	1
6/5/2017 13:00	71	0	2	0
6/5/2017 14:00	53	0	0	0
6/5/2017 15:00	52	0	0	0
6/5/2017 16:00	48	0	5	0
6/5/2017 17:00	55	0	3	0
6/5/2017 18:00	59	0	4	0

Table 8.7 Comparison between multiple-hour RPC and single-hour optimization

	Multiple-hour RPC	Multiple single-hour optimization runs
Number of controls moved	138	248
Number of same shunt switching more than once	None	>40
Maximum number of unsolved constraints in an hour	10	10

The multiple-hour RPC was running for a look-up period of 6 hours starting at 16: 00 on March 17, 2018. In this case, the load was relatively low, and the wind generation changed by a large amount, resulting in a huge number of initial high voltage violations. With the available controls, the multiple-hour optimization brought the violations to a much lower number. It is seen that the multiple RPC is very effective in reducing the number of base case voltage violations, thus improving the power grid security especially when the operation conditions are varying over the next few hours.

7   Conclusions

This chapter presents an implementation of linear approach for reactive scheduling in a multiple-period horizon. This method minimizes the number of switching actions on the whole control space and also minimizes the number of switching on

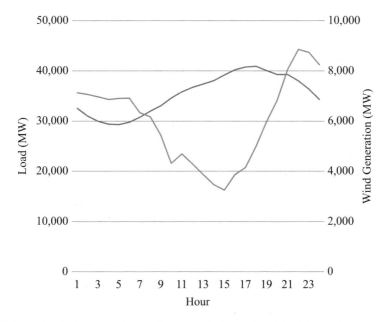

Fig. 8.4 Load and wind generation profile on March 17, 2018 (red, load; blue, wind generation)

individual capacitors. This method also uses node breaker model to accommodate special switching sequences of some shunt devices. The proposed method was tested on ERCOT system and has demonstrated its advantages over current practices (no optimization is performed) and single-period optimization as it can effectively reduce the number of voltage violations for both base cases and contingencies.

References

1. www.ercot.com
2. X. Luo, and O. Obadina, "Security assessment and enhancement in real-time operations of ERCOT nodal electricity market," *IEEE PES General Meeting*, 2010, pp. 1–7.
3. Daniel, J., C. Han, S. Hutchinson, R. Koessler, D. Martin, G. Shen, and W. Wong, ERCOT CREZ reactive power compensation study, ABB Inc., Power Systems Division, Grid Systems Consulting, 2010.
4. Mary B. Cain, Richard P. O'Neill, Anya Castillo, "History of optimal power flow and formulations," FERC staff technical paper, December, 2012.
5. Marija D. Ilic, and John Zaborszky, Dynamics and Control of Large Electric Power Systems, New York: Wiley, 2000.
6. Venkataramana Ajjarapu and Colin Christy, "The continuation power flow: a tool for steady state voltage stability analysis," *IEEE Transactions on Power Systems,* 1992, 7, (1), pp. 416–423.
7. Wenjuan Zhang, Fangxing Li, and Leon M. Tolbert. "Review of reactive power planning: objectives, constraints, and algorithms," *IEEE Transactions on Power Systems*, 2007, 22, (4), pp. 2177–2186.

8. Mohseni-Bonab, Seyed Masoud, and Abbas Rabiee. "Optimal reactive power dispatch: a review, and a new stochastic voltage stability constrained multi-objective model at the presence of uncertain wind power generation," *IET Generation, Transmission & Distribution*, 2017, 11, (4), pp. 815–829.

9. H. Sun, Q. Guo, J. Qi, V. Ajjarapu, R. Bravo, J. Chow, Z. Li, R. Moghe, E. Nasr-Azadani, U. Tamrakar, and G.N. Taranto, "Review of challenges and research opportunities for voltage control in smart grids," *IEEE Transactions on Power Systems*, 2019, 34, (4), pp.2790–2801.

10. Venkatasubramanian, Vaithianathan, Javier Guerrero, Jingdong Su, Hong Chun, Xun Zhang, Farrokh Habibi-Ashrafi, Armando Salazar, and Backer Abu-Jaradeh, "Hierarchical two-level voltage controller for large power systems," *IEEE Transactions on Power Systems*, 2015, 31, (1), pp. 397–411.

11. Ajjarapu, Venkataramana, Ping Lin Lau, and Srinivasu Battula, "An optimal reactive power planning strategy against voltage collapse," *IEEE transactions on Power Systems*, 1994, 9, (2), pp. 906–917

12. L. Bruno and V. Ajjarapu, "An approach for real time voltage stability margin control via reactive power reserve sensitivities," *IEEE Transactions on Power Systems*, 2013, 28, (2), pp. 615–625.

13. Feng Dong, Badrul H. Chowdhury, Mariesa L. Crow, and Levent Acar, "Improving voltage stability by reactive power reserve management," *IEEE transactions on Power Systems*, 2005, 20, (1), pp. 338–345.

14. T.J. Bertram, K.D. Demaree, L.C. Dangelmaier, "An integrated package for real-time security enhancement," *IEEE Transactions on Power Systems*, 1990, 5, (2), pp. 592–600.

15. T. Gomez, I.J. Perez-Arriaga, J. Lumbreras, and V.M. Parra, "A security-constrained decomposition approach to optimal reactive power planning," *IEEE Transactions on Power Systems*, 1991, 6, (3), pp. 1069–1076.

16. P. Lagonotte, J. C. Sabonnadiere, J. Y. Leost, J. P. Paul, "Structural analysis of the electrical system: Application to secondary voltage control in France", *IEEE Trans. Power Syst.*, 1989, 4, (2), pp. 479–486.

17. Hongbin Sun, Qinglai Guo, Boming Zhang, Wenchuan Wu, and Bin Wang, "An adaptive zone-division-based automatic voltage control system with applications in China," *IEEE Transactions on Power Systems*, 2012, 28, (2), pp. 1816–1828.

18. M. Ilic-Spong, J. Christensen, and K.L. Eichorn, "Secondary voltage control using pilot point information," *IEEE Transactions on Power Systems*, 1988, 3, (2), pp.660–668.

19. Qinglai Guo, Hongbin Sun, Mingye Zhang, Jianzhong Tong, Boming Zhang, and Bin Wang, "Optimal voltage control of PJM smart transmission grid: Study, implementation, and evaluation," *IEEE Transactions on Smart Grid*, 2013, 4, (3), pp. 1665–1674.

20. Tao Ding, Rui Bo, Hongbin Sun, Fangxing Li, and Qinglai Guo, "A robust two-level coordinated static voltage security region for centrally integrated wind farms," *IEEE Transactions on Smart Grid*, 2015, 7, (1), pp. 460–470.

21. Lu, Yidan, A Wide Area Hierarchical Voltage Control for Systems with High Wind Penetration and an HVDC Overlay, 2017.

22. Tao Ding, Shiyu Liu, Zhongyu Wu, and Zhaohong Bie, "Sensitivity-based relaxation and decomposition method to dynamic reactive power optimisation considering DGs in active distribution networks," *IET Generation, Transmission & Distribution*, 2017, 11, (1), pp. 37–48.

23. Hanchen Xu, Alejandro Dominguez-Garcia, and Peter W. Sauer, "Optimal tap setting of voltage regulation transformers using batch reinforcement learning," *IEEE Transactions on Power Systems*, 2019, 35, (3), pp. 1990–2001.

24. Diao, Ruisheng, Zhiwei Wang, Di Shi, Qianyun Chang, Jiajun Duan, and Xiaohu Zhang. "Autonomous voltage control for grid operation using deep reinforcement learning," arXiv preprint arXiv:1904.10597, 2019.

25. Le Xie, Diran Obadina, Xinbo Geng, Dongqi Wu, Look-ahead Optimal Reactive Power Dispatch, 2018.

26. Abbas Rabiee, and Mostafa Parniani, "Voltage security constrained multi-period optimal reactive power flow using benders and optimality condition decompositions," *IEEE Transactions on Power Systems*, 2012, 28, (2), pp. 696–708.

Chapter 9
Renewable Forecast

1 Introduction

In the electricity grid at any moment, balance must be maintained between electricity consumption and generation—otherwise, the disruptions of supply or involuntary load shedding may occur. As wind and solar generation is non-dispatchable, variable, and intermittent, in contrast to conventional generation systems, the fluctuations of renewable generation resources require power substitution from other dispatchable sources. However, doing so requires a small percentage of the overall generation capacity reserved, and it may also take some time to bring these resources online. To assist this, renewable energy forecasting is considered as one of the most effective ways to integrate renewable resources reliably and efficiently.

Renewable energy forecasting may be considered at different time scales, depending on the intended application:

Very-short-term forecasts (from seconds up to minutes) are used for the real-time control and electrical grid management, as well as for market clearing.
Short-term forecasts (from 30 minutes up to hours) are used for dispatch planning and intelligent load shedding decisions.
Medium-term/long-term forecasts (from a day up to a week or even a year) are used for long-term planning (to schedule the maintenance or unit commitment and optimize the cost of operation).

Renewable energy forecasting in its nascent stage began in the early years of the twenty-first century. The most prevailing method to predict the renewable power production potential is based on numerical weather prediction (NWP), which uses mathematical models of the atmosphere and oceans to predict the weather based on current weather conditions.

This chapter discusses the wind and solar forecasting system deployed at ERCOT to help to maintain the grid reliability.

© The Author(s), under exclusive license to Springer Nature Switzerland AG 2023   243
P. Du, *Renewable Energy Integration for Bulk Power Systems*, Power Electronics
and Power Systems, https://doi.org/10.1007/978-3-031-28639-1_9

Table 9.1 Specifications of wind forecasting products and services

	STWPF and WGRPP	Extreme weather forecast	Intra-hour wind forecast
Time resolution	1 hour	1 hour	5 minutes
Forecasting horizon	168 hours	168 hours	2 hours
Delivery	ERCOT total/wind regions/WGRs	ERCOT total/wind regions/WGRs	ERCOT total/wind regions/WGRs

2 Wind Forecasting System

Per the Nodal Protocols, ERCOT is required to produce forecasts of renewable production potential (RPP) for wind power generation resources (WGR) and photovoltaic generation resource (PVGR). This forecasting is essential to the reliable integration of a large amount of renewable resources into ERCOT.

Since 2010, ERCOT has contracted one wind forecasting service provider to provide wind generation forecast. It originally consisted of hourly short-term wind power forecast (STWPF[1]) and wind-powered generation resource production potential (WGRPP) for each wind generation resource (WGR) and the ERCOT system aggregation for the next 168 hours. In addition to hourly wind forecast, the wind forecasting products and services also were expanded to include intra-hour wind forecast and extreme weather forecast. The specifications for wind forecasting products and services are given in Table 9.1.

In 2018, ERCOT integrated a second wind forecasting service provider due to the critical role of wind forecasting in the grid operations as the amount of wind installed capacity continues to grow. Adding another wind forecasting service provides ERCOT's control room with an alternative wind forecast to enhance situational awareness, increase resiliency for the provision of this critical information, and offer new capabilities on forecasting for extreme weather conditions.

2.1 Wind Forecasting System Overview

The primary purpose of the wind forecasting system is to produce forecasts for wind-powered generation resources (WGR) so as to know how much the wind power production will be anticipated for the future hours, which is considered as an important input into the day-ahead reliability unit commitment (DRUC) and hour-ahead reliability unit commitment (HRUC).

---

[1] STWPF and WGRPP represent the 50th and 80th percentiles of the wind forecast, respectively.

In addition to STWPF and WGRPP, another product delivered from the wind forecasting system is to produce an extreme weather for the next 168 hours so as to predict (1) the risk of incoming extreme weather events and (2) the impact of extreme weather events over the wind generation. This extreme weather forecast includes a worst scenario with all of wind turbines out of the service caused by extreme weather conditions and an expected scenario with 50% of wind turbines impacted offline. The derating of wind farms resulting from the impact of extreme weather events is created on an hourly basis for the next 168 hours, and the delivery time of the icing forecast is synchronized with that of STWPF and WGRPP. A process is also provided to allow the ERCOT users to manually include or exclude the extreme weather's impact into the STWPF delivered. A comparison between normal and extreme weather forecast is given in Table 9.2.

Upon the receipt of the latest available wind forecasts (STWPF and extreme weather forecast), operators will choose one forecast at their discretion for the downstream applications according to the incoming weather conditions. ERCOT will automatically use this forecast to update current operation plan (COP) of WGRs for the next 168 hours. If qualified scheduling entities (QSEs) need to account for an outage not captured in the latest wind forecast, QSEs can sequentially change their COP in such a way that the COP submitted by WGRs should be less than or equal to the latest wind forecast for the corresponding WGRs. This is accomplished before the top of the next hour so that the execution of the next HRUC can utilize the latest wind forecast available for the targeted operating hour. Figure 9.1 shows a timeline for wind forecast delivery, selection, and COP update.

Table 9.2 Comparison between normal and extreme weather forecast

Normal weather forecast	STWPF/ WGRPP	Exclude the impact of icing caused by extreme weather conditions
Extreme weather forecast	Worst icing scenario	All of wind turbines impacted by extreme weather conditions are out of the service
	More likely scenario	50% of wind turbines impacted by extreme weather conditions are offline

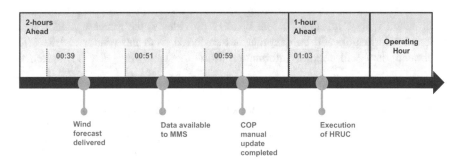

Fig. 9.1 Timeline for wind forecast delivery, selection, and COP update

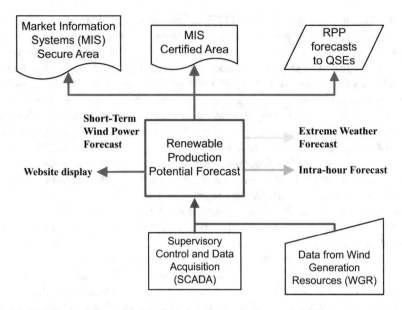

Fig. 9.2 Data flow and dependencies of wind power production potential forecast

The third wind forecasting product delivered to ERCOT is an intra-hour wind forecast produced for every 5 minutes of the next 2 hours. The ERCOT-level wind power production forecast for the next 5 minutes is incorporated into the generation to be dispatched (GTBD)[2] calculation. Adding the forecasted change in wind generation power potential for the next 5 minutes in GTBD will allow for a more efficient dispatch and a better management of regulation resources.

2.2 Data Flow of Wind Forecasting System

A high-level description of data flow of the wind forecasting system is depicted in Fig. 9.2.

A structure of the wind forecasting system is described as follows. Outbound information will be information that ERCOT provides to the forecast vendor, and inbound information will be data that is transferred from the vendor to ERCOT. Outbound data will require information from the following ERCOT systems:

---

[2]GTBD is used in security-constrained economic dispatch (SCED), and it is based on the sum of expected updated desired base points for all generation resources plus the amount of regulation service deployed at the time of the determination of GTBD for each execution of SCED. The other GTBD inputs include forecasted system load change for the next 5 minutes, forecasted change in output of WGRs during the next 5 minutes, and an adjustment to correct load for current frequency that is different from the scheduled frequency.

SCADA[3] subsystem, outage scheduler subsystem, registration subsystem data, and EMS Oracle database. This is accomplished using a communication server.

On the outbound, the wind forecasting system will consolidate the required data from the ERCOT systems and then deliver this input information to the vendor server. The vendor will then create a short-term (168-hour) forecast. On the inbound, the vendor communication server will deliver the forecast to ERCOT. This information will be provided to the QSE via the messaging system, posted on the market information system.

The wind forecasting system will provide a user interface to display the forecast and performance data. This ERCOT operator user interface will support large geographic displays as well as database-driven tabular displays used to monitor the process of preparing, transferring, and receiving forecast information. File transfer failures will issue alarms to the EMS support teams. Typically, the wind forecasting system requires a high level of availability with no schedule downtime and is expected to be running 24 hours a day, 365 days a year.

2.3   Input Data for Wind Forecasting System

WGR entities must provide registration, outage scheduler, and real-time telemetry data to be used as an input to the wind generation forecast models. ERCOT will receive the SCADA information for the WGR from the WGRs QSE as well as the registration and outage scheduling information.

Input data consists of operational data, registration data, and outage data. All the operational input data will be telemetered via SCADA from the WGR to its QSE and then to ERCOT via Inter-Control Center Communications Protocol (ICCP). This data is then inserted into a database at ERCOT before being sent to the wind forecasting service each 5 minutes. This SCADA data will belong to the current 5-minute time interval, and any integration of wind speed or MW output will be done in the ERCOT SCADA system. Static data will be sent to the wind forecasting system through a XML file once every week to reflect registration changes, which are made via market participant generation asset registration forms (GARF). The expectation is that current and new wind generation facilities will have the required meteorological data from meteorological towers located at the site. All operational information will be sent to the forecasting vendor each 5 minutes.

The data sent to the forecasting vendor includes:
Registration data:

1. Resource name
2. Resource-QSE mapping
3. Resource parameters max and min reasonability limits (units in MW)

---

[3] SCADA: supervisory control and data acquisition

4. Location of wind farm latitude and longitude for the center point of wind farm (units in decimal degrees)
5. Location of the meteorological tower latitude and longitude (units in decimal degrees)
6. Type (manufacturer/model) and number of turbines
7. Turbine hub height(s) above ground level with associated number of turbines (units in meters)
8. Meteorological tower height
9. Manufacturer's power curve (capability curve)
10. Resource commercial operation date

Outage scheduler data:

1. Scheduled outages/de-rates of wind farms

SCADA data:

1. Resource (wind farm) status online/offline with date/time (number of online/offline/unknown wind turbines)
2. MW output of wind farm with date/time
3. HSL of wind farm with date/time
4. Curtailment flag
5. Wind speed (units in MPH)
6. Wind direction from one meteorological tower with date/time
7. Temperature (units in °C)
8. Barometric pressure at 2 m above ground level on the meteorological tower (units hpa (mb))

2.4 Design Approach of Wind Forecasting System

The objective of the forecasting service system is to create a forecast service which will provide ERCOT with the most accurate and reliable wind forecasts possible. The hourly forecasts will be updated hourly and cover 1–168 hours from the forecast update time for both normal and extreme weather conditions. The intra-hour wind forecast will be updated each 5 minutes and cover the next 2 hours. The service will be developed to provide an extremely high level of reliability with over 99.8% of the forecasts delivered on time. ERCOT has contracted to two wind forecasting vendors for their forecasting service given their extensive experience in wind power forecasting and research.

ERCOT will place all EMS real-time and outage scheduler data each 5 minutes in a communication server. The server will deliver the input data to each vendor each 5 minutes and currently, and both two vendors support web services. Since the source system for outage scheduler data is an Oracle database, the wind forecasting system will pull this data from the source database to the staging database. EMS real-time SCADA data is stored in General Electric's proprietary database, Habitat.

Therefore, General Electric's eTerra Archive software product will be used to extract the SCADA data from the proprietary format and transfer it to the Oracle staging database. Once all of the data has been organized and consolidated, the forecasting system will transfer this data to the communication server for communication to the servers at each wind forecasting service provider (WFSP) site. The WFSPs will then create and provide a 168-hour rolling WGRPP, STWPF, and the synchronized extreme weather forecast to ERCOT before the next operating hour. Regarding the intra-hour forecasting, the WFSPs will send ERCOT the 5-minute forecast quantities about 1 or 2 minutes prior to the next SCED execution.

The registration data is automatically transferred to the WFSPs through a separate XML file, which contains the changes expected to occur in the near future (e.g., a new site or a QSE change for a site).

The forecasting system developed by the WFSPs consists of several elements. Each component of the system will have at least one backup. A suite of physical weather models will be used to forecast conditions over the entire state of Texas for the 168-hour forecasting period. The models provide the capability to make a forecast even when ERCOT-provided WGR data is unavailable. A second component of the wind forecasting system will be a suite of statistical models. The statistical models use plant data to eliminate bias in the physical model forecasts and to provide information about recent trends in the power production which are valuable predictors in the first few hours of the forecast period. The statistical models will be able to function with only historical data although current data will significantly improve the first few hours of the forecast. An ensemble compositing algorithm will be used to produce a final forecast from the ensemble of statistical model forecasts. For each WFSP, a delivery control system will be put in place to ensure that the forecast is sent from one of several redundant systems. The forecast will be delivered via the Internet. To ensure reliability, two connections to the Internet for each WFSP will be maintained. The forecast systems will be located behind a firewall in a server room monitored for unauthorized access and environmental conditions (temperature, humidity, liquid water, etc.).

All processing to create the forecast will be done at the WFSP site. Processing of the physics-based models used in making the forecasts will occur separately from the running of the statistical models and other forecast processing as it will take substantially more processing time than other forecast system components. The physics-based models will be updated 2–24 times per day with different update times and frequencies for different models, and the other components of the system will use the most recent available physics-based model data. It is anticipated that the forecast system will require 15 minutes or less to generate a forecast from the time when new ERCOT data is received by the WFSP.

All processing will be done at the WFSP's facilities. ERCOT will provide a method of sending web service information containing WGR data to the WFSP and a method of receiving web service information containing the forecast data from the WFSP. Transmission of the data will be via the Internet over a secure channel using digital certificates as either secure FTP or web services over HTTPS.

The WFSP will deliver forecast data to ERCOT via the Internet over a secure channel as well. The WFSP will use web services over HTTPS with the use of digital certificates.

3   Solar Forecasting System and Forecast Errors

This section provides a detailed overview of the technology used in solar forecasting system and the analysis of solar forecast errors.

3.1   Solar Forecasting System

Solar forecasting system is a suite of modules and tools designed specifically to meet the renewable energy industry's need for accurate plant-level and regional-scale weather and power production forecasts from 5 minutes to months in advance. It contains six basic elements: (1) a set of physics-based numerical weather prediction (NWP) models, (2) a suite of advanced statistical prediction techniques, (3) a set of cloud tracking algorithms, (4) a set of plant output models, (5) an ensemble optimization algorithm, and (6) a mechanism to deliver forecasts.

3.1.1   NWP Data

The solar forecasting system uses grid-point output from global-scale and, where available, regional-scale NWP models executed at government-sponsored forecast centers across the world. Table 9.3 summarizes the NWP models that are available to be used for ERCOT's solar forecast system and the government center from which it originates.

3.1.2   Machine Learning Algorithms Used by Solar Forecasting System

The solar forecasting system runs a suite of machine learning algorithms at the frequency that forecast updates are needed by the end user (e.g., every 5 minutes if a forecast update is needed every 5 minutes). Inputs to the machine learning algorithms include (1) the most recent output from all NWP models in the NWP ensemble, (2) any meteorological measurements from publicly available sensors, (3) satellite cloud information, and (4) on-site wind and solar plant measurements. The on-site plant data typically consists of generation and facility availability, as well as meteorological variables.

Table 9.3 Summary of government-sponsored NWP models

Government center	NWP model	Forecast duration	Model update frequency
National Centers for Environmental Prediction (NCEP), US	Global Forecast System (GFS)	384 h	4×/day
NCEP	North American Mesoscale Model (NAM)	84 h	4×/day
NCEP	Rapid Refresh (RAP)	48 h 4×/day; otherwise, 21 h	Hourly
NCEP	High-Resolution Rapid Refresh (HRRR)	48 h 4×/day; otherwise, 18 h	Hourly
NCEP	Global Ensemble Forecast System (GEFS)	384 h	4×/day
Environment Canada (EC)	Global Deterministic Prediction System (GDPS)	240 h	2×/day
EC	Canadian Ensemble Forecasts (MSC)	384 h	2×/day
European Centre for Medium-Range Forecasts (ECMWF)	ECMWF Global Model	240 h	2×/day
United Kingdom Met Office (UKMET)	UKMET Global Model	144 h	2×/day
Deutscher Wetterdienst (DWD, the German Meteorological Office)	Icosahedral Nonhydrostatic (ICON) Global Model	180 h	4×/day

The forecasting system has an extensive set of quality control (QC) software to filter erroneous or unrepresentative data from any sources used in the forecast system. The QC system uses a wide range of tools from simple range checking of individual variables to advanced statistical procedures that identify subtle problems with input data or output forecasts in real time. The QC step is important because it ensures that a representative sample of data is available for training the statistical models.

The primary role of the machine learning algorithms is to derive empirical predictive relationships from a historical data set that includes NWP model output, satellite cloud information, and local generation and meteorological measurements. They are also used to reduce NWP model biases. A real-time feed of on-site data into the machine learning algorithms is critical in adjusting NWP model data for changing on-site weather conditions. Any of these machine learning algorithms can be used with the output from one or more members of the NWP ensemble. Similar to the NWP ensemble, the composition of the statistical model ensemble is periodically reviewed to ensure the latest and most accurate forecast technology is being employed.

3.1.3 Satellite Cloud Tracking Algorithms

The forecasting system has developed, tested, and implemented a solar resource estimation method using satellite-based imagery to estimate and predict irradiance. The irradiance estimation method processes raw satellite imagery from the Geostationary Operational Environmental Satellite (GOES) satellites operated by NOAA's National Environmental Satellite, Data, and Information Service (NESDIS).

The solar forecasting system uses a clear-sky model that accounts for solar elevation angle, atmospheric turbidity, and terrain elevation to compute irradiance expected with no cloud cover. The GOES data are used to adjust the raw clear-sky irradiance estimate at grid points using a cloud index derived from visible brightness data and calibrated with ground-based irradiance measurements, such as those provided from a plant (in situ), publicly available observations, or UL proprietary data. The ground-based measurements are also used to perform a bias correction to account for local factors (e.g., local variations in the steepness and direction of the terrain slope). The result is a gridded 0.5 km resolution map of fraction of clear-sky irradiance roughly every 5 minutes, which is used to estimate actual irradiance (W/m^2). Figure 9.3 shows the correlation and mean absolute error (MAE) of the solar forecasting system at multiple climate stations around the contiguous united

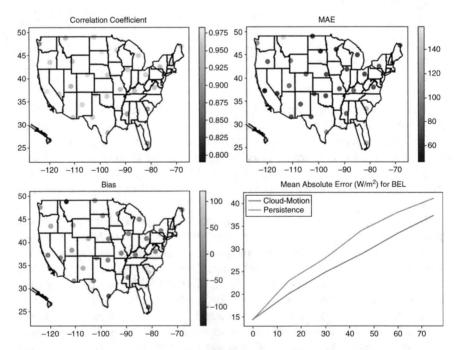

Fig. 9.3 Correlation coefficient (top left), mean absolute error (W/m^2, top right), and bias (W/m^2, bottom left) of satellite-estimated GHI vs. surface observations of GHI at US Climate Reference Network Stations. (Lower right) Cloud-motion based forecast mean absolute error (W/m^2) versus persistence of clear-sky GHI forecast for a single location in New York State

stations. Correlations and errors are shown for 15-minute averages to account for the difference between spatial scale of observation (point measurements from the ground vs. 0.5 km × 0.5 km areal averages from space). Correlations are typically well above 0.93, with MAE typically around 60 W/m^2.

A cloud-advection model is then used to extrapolate the motion of cloud features diagnosed in visible satellite imagery. The solar forecasting system uses the changes in a quantity called the clear-sky index (CSI) from successive visible satellite images to forecast future CSI. The CSI fields are created using a model that uses high-resolution gridded visible satellite cloud cover as input and averages it to a series of progressively lower-resolution grids. The solar forecast estimates a coarse motion field based on two consecutive low-resolution images and then refines this field by computing corrections at successively higher resolutions. By starting at low resolution, it reduces the likelihood of errors from mapping a fine-scale feature at Time A to a different feature at Time B. It also produces a smooth estimate of the motion field at each pixel, allowing different features to move at different speeds and directions. Once an accurate estimate of the motion field is obtained, the most recent CSI image is projected forward in time over the next 6-hour period to create irradiance and power production forecasts at each solar power generating site. It also uses non-visible satellite data (infrared) in the solar irradiance estimation technique to update pre-dawn forecasts.

3.1.4 Plant Output Models

Plant output models generally take two forms and are used to transform predictions of meteorological variables into predictions of generation. The first is an explicit plant output model in which NWP and statistical models are used to create predictions of meteorological variables. The predicted meteorological variables are then fed into an equation that relates meteorological variables to generation. The equation may be either the manufacturer's power curve information (typically used when a wind/solar plant has just started generating) or an equation developed from the observed on-site generation and meteorological data. The second form of plant output model is an implicit one, in which statistical models are used to directly relate NWP data to the observed on-site generation.

The solar forecasting system will use both types of plant output models. The explicit method is employed for new facilities with a limited sample of observed data, as well as during times of curtailment, regardless of the period of record available. Once enough data are acquired, the implicit method is used in addition to the explicit method. Forecasts from both methods are then used as inputs to the ensemble optimization algorithm, which weights them according to their relative accuracy compared to other methods over the ensemble training period.

3.1.5   Ensemble Optimization Algorithm

An ensemble (set) of forecasts is generated from the combination of NWP models, statistical models, cloud tracking algorithms, and plant output models. All of these forecasts are available as inputs to an ensemble optimization algorithm (EOA). The EOA creates a composite forecast from the set of individual forecasts by weighting them based on past performance and the expected conditions at the location of interest. The weighting is done for each look-ahead interval in the forecast, meaning that the composition of the ensemble and weighting of the individual forecasts used to produce a 5-minute-ahead forecast may be very different from the composition and weighting of a day-ahead forecast.

If a customer requests probabilistic forecast information, the individual model forecasts as well as components of the raw NWP output are used to train a machine learning-based forecast of cumulative probability distribution. This probability distribution is used to construct probability of exceedance (POE) forecasts around the composite forecast value for each interval in the forecast period. Large differences between the POE forecast values and the composite forecast value indicate the potential for variability and uncertainty in the forecast, while small differences convey greater certainty. The solar forecasting system can generate the probabilistic forecast data to deliver any number of POE values.

3.1.6   Solar Forecast Delivery Mechanism

Once the forecast has been created in the desired format, the final step is to communicate the forecast to the client. The solar forecasting system at ERCOT is based on ensemble forecast system, which is highly robust and fully redundant, capable of achieving more than 99.9% availability. It uses NWP models from multiple government centers on different continents as inputs into an array of statistical forecasting models. If the transfer of NWP data from one or even two government centers is disrupted, a high-quality composite forecast will still be generated. The forecast system also has the capability of using previous NWP model cycles if the most recent one is unavailable.

3.2   Solar Forecast Error Analysis

Solar generation resources are one of the fastest-growing resources installed globally to provide an alternative to conventional energy sources. On the other hand, as more utility-scale solar photovoltaics (PV) systems are built, their impacts over the grid operations and reliability must be thoroughly examined, especially due to their variations and intermittence [1]. This necessitates the characterization of the solar forecasting errors (SFE) so that the realistic SFEs can be replicated when studying the impact of solar PV systems over the power grid. However, in contrast to the wind forecasting errors [2, 3], the characterization of SFEs has not been studied except in [4]. Different from the work in [4], this section utilizes a t-copular to model dependent extreme value for the temporal correlations between SFEs.

3.2.1   Source of Data

ERCOT works with a third-party forecasting service provider to run an ensemble-based numerical weather prediction (NWP) model, which produces solar forecast for the next 168 hours for each utility-scale solar farm. The solar forecasting data and solar power production during a 3-year period (2016–2018) were collected for the analysis. The data is aggregated over the whole ERCOT system and normalized with respect to the installation capacity at the time when the data was produced. The data set was also truncated by removing those night hours.

The SFE is calculated as the difference between the solar power production and the forecasted value, which is

$$e_{t,t-h} = p_t - \tilde{p}_{t-h} \tag{9.1}$$

where p_t is the actual solar power production at hour t, $\tilde{p}_{t-h}$ is the solar forecast produced at hour $t-h$ for hour t, and $e_{t,\,t-h}$ is the h-hour-ahead SFE for hour t.

Since ERCOT is executing an hourly reliability unit commitment (HRUC) market to evaluate the resource adequacy for the future operation hours, the leading time from 3 hours to 1 hour prior to the operational hour is critical to commit additional resources if there is need. Given this, the analysis conducted in this chapter is focused on 1-hour-, 2-hour-, and 3-hour-ahead SFE, $e_{t,\,t-h}$ ($h = 1, 2, 3$). Figure 9.4 shows the joint distribution of 1-hour-ahead solar forecast, $\tilde{p}_{t-1}$, and actual power production, p_t, and their empirical marginal probability distribution function (PDF).

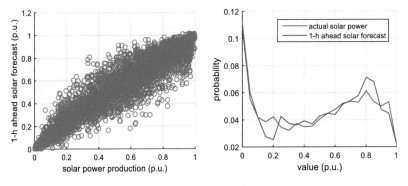

Fig. 9.4 Joint distribution between 1-hour-ahead solar forecast and actual power production (left) and marginal probability distribution (right)

The major factors influencing the solar power production include the hour of day, season, cloud movement, and back-panel temperature.

3.2.2 Probability Distribution Function of Solar Forecast Errors

An understanding of the PDF of SFEs can facilitate the reproduction of SFEs in a variety of solar integration studies. To analyze the PDF of SFEs, a similar methodology is used where the forecasted solar power time series are sorted into power classes or bins (index, i, set I) according to the forecasted value, $\tilde{p}_{t-h}$ [2, 3]. Then SFEs are then calculated on each time step within a specific power bin i. The conditional prediction error for a power bin i is given by

$$\varepsilon_{t,t-h,i} = p_t - \tilde{p}_{t-h} (L_i \leq \tilde{p}_{t-h} < U_i) \tag{9.2}$$

Figure 9.5 depicts a joint distribution between 1-hour-ahead solar forecast and the corresponding SFEs. A total of 20 power bins with an equal width are selected. It is shown that 1-hour-ahead solar forecast tends to be under-forecasted when its value is low. On the other hand, the NWP model likely produces an over-estimation of solar power production when it is close to the full capacity. An over-estimation of solar power represents a reliability concern for the grid operations, while an under-forecast may decrease the market efficiency.

For each power bin i, the histogram of the error $\varepsilon_{t,\ t-h,\ i}$ can be obtained, and an appropriate distribution can be used to describe it. Among 17 distributions, the extreme value, logistic, and t-location-scale distributions were found to better approximate 1-hour-ahead SFEs. The parameters of these distributions are estimated by fitting the distribution to the SFE data for each power bin using maximum

Fig. 9.5 Joint distribution between 1-hour-ahead solar forecast and corresponding SFE

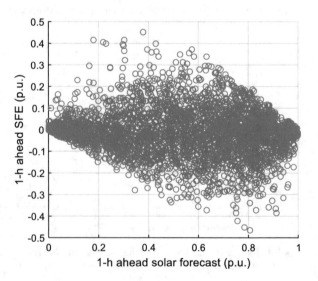

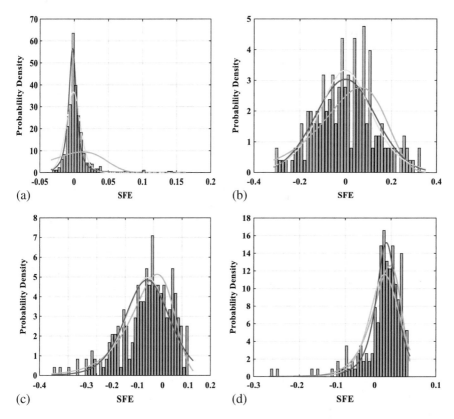

Fig. 9.6 PDF of 1-hour-ahead SFE for selected power bins (blue, empirical; green, extreme value; red, t-location scale; orange, logistic). (**a**) Power bin #1 $(0 \leq \tilde{p}_{t-1} < 0.05)$. (**b**) Power bin #11 $(0.5 \leq \tilde{p}_{t-1} < 0.55)$. (**c**) Power bin #18 $(0.85 \leq \tilde{p}_{t-1} < 0.9)$. (**d**) power bin #20 $(0.95 \leq \tilde{p}_{t-1} < 1)$

likelihood estimation. Some examples are given in Fig. 9.6. For power bins located close to both ends of the solar power production range, t-location-scale distribution is the best fit to approximate the empirical histogram. In the middle range of the solar power production, both t-location-scale and logistic distribution yield a similar performance.

The PDF of the t-location-scale distribution is

$$
\frac{\Gamma\left(\frac{\nu+1}{2}\right)}{\sigma\sqrt{\nu\pi}\Gamma\left(\frac{\nu}{2}\right)} \left[\frac{\nu + \left(\frac{x-\mu}{\sigma}\right)^2}{\nu} \right]^{-\left(\frac{\nu+1}{2}\right)}
\tag{9.3}
$$

where $\Gamma(\cdot)$ is the gamma function, μ is the location parameter, σ is the scale parameter, and ν is the shape parameter.

After the estimation process, for each power bin, the parameter pair $\{\mu, \nu\}$ of t-location-scale distribution can be derived, which is provided in Table 9.4.

Table 9.4 Parameters of t-location-scale distribution

Power bin #	Location	Scale	Shape
1	0.00	0.01	1.8
2	0.00	0.02	3.6
3	−0.01	0.04	4.1
4	−0.02	0.06	3.3
5	−0.02	0.05	2.8
6	−0.02	0.07	4.8
7	−0.01	0.10	8.5
8	−0.01	0.12	39.9
9	0.00	0.14	3,137,036
10	0.00	0.13	3,458,308
11	0.00	0.13	2,305,315
12	−0.02	0.13	24.8
13	−0.03	0.14	2,610,932
14	−0.01	0.13	3,877,903
15	−0.01	0.11	60.2
16	−0.02	0.11	12.6
17	−0.04	0.10	16.7
18	−0.02	0.08	9.0
19	−0.04	0.06	4.2
20	−0.01	0.02	2.3

To obtain an unconditional PDF for SFE, the PDFs of each bin $f(\varepsilon_{t,\ t-h,\ i})$ are combined via the empirical probability that a solar forecast value will be within the range of that bin, $f(b_i)$, which is

$$f = \sum_{i=1}^{m} f(\varepsilon_{t,t-h,i}) f(b_i) \tag{9.4}$$

where m is the number of power bin ($m = 20$).

Figure 9.7 shows skewness and kurtosis of 1-hour-ahead SFEs for power bins #1–#20. The most right point with a kurtosis of 36 and a skewness of 4.7 corresponds to power bin #1. As the number of power bin increases, the skewness decreases. At the middle range of power bins, the associated kurtosis and skewness are close to zero as the shape of PDF of SFEs is similar to a normal distribution. However, as 1-hour-ahead solar forecast value approaches 0 per unit, the kurtosis increases, and the skewness even becomes negative.

3.2.3 Temporal Correlation of SFE

As the time moves toward the operation hour, the NWP model is updated. However, due to a weather momentum, the change in the solar forecast value may be small or modest, thus leading to a correlation among 3-hour-ahead, 2-hour-ahead, and

Fig. 9.7 Skewness and kurtosis of SFE for power bins #1–#20

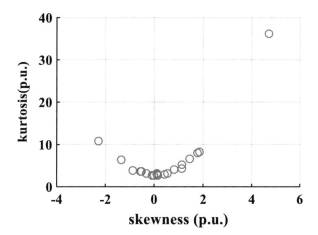

1-hour-ahead SFEs. This correlation can be characterized by a copular function [5, 6].

A d-dimensional copula C is a d-dimensional distribution function (DF) with standard uniform marginal distributions. Sklar's theorem states that every DF F with margins $F_1, \ldots, F_d$ can be written as

$$F(x_1, \cdots x_d) = C(F_1(x_1), \cdots F_d(x_d)) \tag{9.5}$$

In the case when marginal DFs are continuous, the copular C of their joint DF can be evaluated as

$$C(u_1, \cdots u) = F\left(F_1^{-1}(u_1), \cdots F_1^{-1}(u_d)\right) \tag{9.6}$$

Recent work found that the empirical fit of t copula is superior to that of Gaussian copular due to its ability to capture better the phenomenon of dependent extreme value. The density of the t copula has the form

$$c_{w,P(u)}^t = \frac{f_{v,P}\left(t_w^{-1}(u_1), \cdots, t_w^{-1}(u_d)\right)}{\prod_{i=1}^{d} f_{w,P}\left(t_w^{-1}(u_i)\right)} \tag{9.7}$$

$$u \in (0,1)^d \tag{9.8}$$

where u is uniformly distributed between 0 and 1, $f_{w,\ P}$ is the joint density of a $t_d(w, \mathbf{0}, P)$-distributed random vector, and f_w is the density of the univariate standard t-distribution with w degrees of freedom.

The procedures to estimate the parameters of t copula are described as follows. First, transform $e_{t,\ t-h}$ ($h = 1, \cdots, 3$) for different operation hours into uniform distribution records, $\widehat{e}_{t,t-h}$, using their own cumulative distribution function (CDF), respectively. For example, the CDF of 1-hour-ahead, 2-hour-ahead, and 3-hour-ahead SFEs for hour ending 17 (HE 17) is given in Fig. 9.8. Second, find the t copula

Fig. 9.8 CDF of 1-hour-
ahead, 2-hour-ahead, and
3-hour-ahead SFE for hour
ending 17 (HE 17)

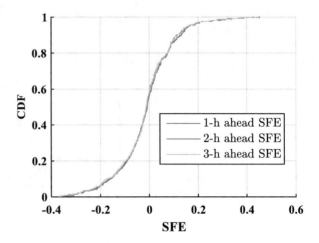

Table 9.5 Estimated the matrix of linear correlation parameters for a t copula for HE 17

	1-h-ahead SFE	2-h-ahead SFE	3-h-ahead SFE
1-h-ahead SFE	1.000	0.993	0.991
2-h-ahead SFE	0.993	1.000	0.995
3-h-ahead SFE	0.991	0.995	1.000

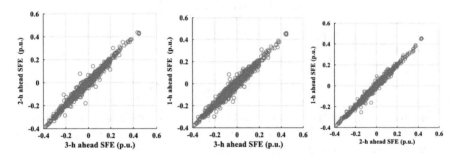

Fig. 9.9 Joint distribution of 3-hour-ahead, 2-hour-ahead, and 1-hour-ahead SFEs for HE 17 (blue, simulated; red, empirical)

by fitting the joint PDF of $\widehat{e}_{t,t-h}$ ($h = 1, \cdots, 3$). Third, reproduce the forecast errors by using estimated t copula to join the collection of univariate DFs in the uniform distribution to validate the results.

After the above procedures are applied to each operation hour, HE 17 is taken as an illustrative example. The estimate of the matrix of linear correlation parameters for a t copula, P, for HE 17 is shown in Table 9.5, and the corresponding degree of freedom parameter, w, is estimated to 1.03. Using estimated parameter in Table 9.5, the SFEs can be reproduced as shown in Fig. 9.9. Compared to empirical data, the simulated data preserves the correlation between solar forecast with different leading times. Figure 9.10 also shows linear dependence between those SFEs analyzed,

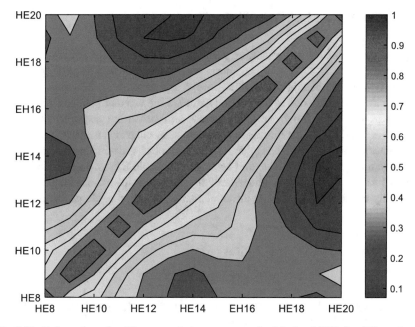

Fig. 9.10 Estimated matrix of linear correlation parameters for 1-h-ahead SFEs for different hours

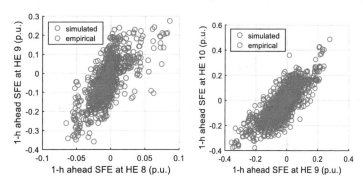

Fig. 9.11 Correlation between 1-hour-ahead SFEs for HE 8, HE 9, and HE 10

which is consistent with the analysis where the off-diagonal element in the estimated correlation matric, P, in Table 9.5 is close to 1.0. This implies that the accuracy of solar forecast for an operation hour will not be able to improve significantly as the leading time of the solar forecast becomes shorter.

The similar procedures are applied to investigate the correlation between SFEs for different operational hours but with a same leading time. Figure 9.10 shows the estimated matrix of linear correlation parameters for a t copula, P, for 1-hour-ahead SFEs and the degree of freedom parameter, w, is estimated to 9.68. The re-constructed 1-hour-ahead SFEs for HE 8, HE 9, and HE 10 are shown in Fig. 9.11, matching well with the empirical data.

4   Summary and Conclusions

Improving wind and solar forecasting performance is one of most effective means to assist in maintaining a balancing operation for a power grid. It is an evolving area as more advanced models and techniques are being developed. However, the challenges still exist today due to the volatile nature of the weather phenomena. Utilities or system operators move toward a practice of procuring multiple vendors to produce a wind and solar forecasting service. This chapter describes in depth the experiences which ERCOT has regarding how to set up the wind and solar forecasting system to produce a reliable and accurate renewable forecasting. Meanwhile, it presents a study performed to understand the characteristics of solar forecast errors, especially the correlations among different sources or extremely large forecast errors ("outliers").

References

1. Du, Pengwei, Ross Baldick, and Aidan Tuohy. "Integration of Large-Scale Renewable Energy into Bulk Power Systems," Springer, 2017
2. Hans Bludszuweit, José Antonio Domínguez-Navarro, and Andrés Llombart, "Statistical analysis of wind power forecast error," IEEE Transactions on Power Systems, vol. 23, no. 3, 2008, pp: 983–991.
3. Bruninx, Kenneth, and Erik Delarue, "A statistical description of the error on wind power forecasts for probabilistic reserve sizing," IEEE transactions on sustainable energy, vol. 5, no. 3, 2014, pp: 995–1002.
4. Edgar Nuño, Matti Koivisto, Nicolaos A. Cutululis, and Poul Sørensen. "On the simulation of aggregated solar PV forecast errors," IEEE Transactions on Sustainable Energy, vol. 9, no. 4, 2018, pp: 1889–1898.
5. George Papaefthymiou, and Dorota Kurowicka, "Using copulas for modeling stochastic dependence in power system uncertainty analysis," IEEE Transactions on Power Systems, vol. 24, no. 1, 2008, pp: 40–49.
6. Zhang, Ning, Chongqing Kang, Qing Xia, and Ji Liang. "Modeling conditional forecast error for wind power in generation scheduling," IEEE Transactions on Power Systems, vol. 29, no. 3, 2013, pp: 1316–1324.

Chapter 10
Ensemble Machine Learning-Based Wind Forecasting to Combine NWP Output with Data from Weather Stations

1 Introduction

Wind generation resources (WGRs) are the fastest-growing energy sources throughout the world due to their environmental advantages and declining capital cost driven by the technology advancement [1]. For some smaller regions, i.e., Texas and Ireland, WGRs already produced a large share of energy to serve the load. As the ambitious goals for renewable portfolio standards are pursued, an extremely high penetration of WGRs would be reached in a large-scale grid in a foreseen future. In a recent study, a scenario in which wind generation could meet 20% of the US electricity demand by 2030 was explored [2]. In Europe, a target of 20/20/20 was established which requires to increase the renewable energy to 20% of the electrical supply by 2020 [3]. While WGRs are bringing tremendous economic and environmental benefits, their intermittent and variable nature also poses a great challenge to maintaining a balance between generation and demand, especially at a high penetration.

The impacts of a high penetration of WGRs over the grid operations across multiple time scales have been extensively studied [1]. Essentially, to ensure the grid's reliability and security, the fluctuations and uncertainties of wind power generation need to be mitigated to some extent, which could require very costly changes to the grid, i.e., increased reserve [4], more flexibility [5], and fast responsive resources like energy storage [6–7]. Among all of these options proposed to accommodate WGRs in the grid operations, an accurate wind forecasting service is critical and most cost-effective, especially when the penetration of WGRs is high [8].

Wind power forecasting (WPF) can predict future wind generation production for the scheduling and dispatching process from day-ahead to hour-ahead operation. By using wind forecasts, system operators can significantly reduce the uncertainty of wind power output. If both load and wind forecast are accurate, the thermal generation units can then be more economically scheduled and dispatched to serve the load. In a case when the wind power is over-estimated, a power supply shortage or a price spike may be experienced. In contrast, if the wind power is under-estimated, a high operational cost may incur as a result of committing more thermal units than necessary [10–11]. In addition, an accurate wind forecasting can improve the grid reliability and security, which allows more WGRs to be integrated into the grid without compromising the reliability [11].

Unlike the load which is highly correlated with the exogenous variables like temperature and the time of a day, the volatile nature of wind makes it more difficult to predict the wind generation with precision. By estimation, the improvement in the wind forecasting performance could result in the cost saving of about 100 million US dollars annually for a medium-size power grid [2]. For this reason, the transmission system operator (TSO), which is responsible for managing the electricity balance on the grid at any time, is in a great need for a more accurate wind forecasting.

There are two main types of WPF in applications, i.e., point forecasting and probabilistic forecasting [12–13]. Point forecasting essentially produces a single value at given points in time, while probabilistic forecasts can estimate the uncertainties of the forecast errors associated with the wind generation predicted. Since the former has been used widely in practice because of its simplicity, this chapter focuses on how to improve the performance of the point wind forecasting for a TSO.

Extensive works have been conducted in the literature of point wind forecasting. A more detailed review on wind forecasting can be found in [8]. A forecasting model consisting of the Gaussian process (GP) with a novel composite covariance function for wind power forecasting was proposed in [14]. An improvement in wind and solar forecasting has been achieved by exploring information from a grid of numerical weather predictions (NWP) [15]. In [16], a sparsity-controlled vector autoregressive model was introduced to obtain sparse model structures in a wind power forecasting. A novel ensemble short-term wind power forecasting model was proposed to use 52 neural network sub-models and 5 GP sub-models in parallel [17]. In [18], the authors investigated a combination of GP and a NWP model to forecast wind power up to 1 day ahead. While all of these preceding works are effective in improving the accuracy of wind generation forecasting, the spatiotemporal correlation from the surrounding regions of WGRs has not been explicitly considered so that their performance could be limited.

The recent development for point wind forecasting attempts to incorporate the spatiotemporal correlation with the nearby locations in the wind forecasting model. In [19], graph learning-based spatiotemporal analysis was conducted for short-term forecast within a wind farm. However, this is only suitable for a short forecasting horizon as only local turbulence can be captured. The authors use off-site predictors to capture spatiotemporal correlations among geographically distributed wind farms [20]. The support vector regression (SVR) was also applied to use the weather data from the weather stations surrounding one wind farm to produce wind forecasting for this particular site [21]. The limiting capabilities of separable models in capturing the spatiotemporal and temporal correlation patterns in wind power generation motivated the researchers to establish a non-separable, asymmetric model that captures spatiotemporal interactions [22]. In [23], the study was based on the data from a remote weather station to provide a few hours-ahead forecast for the wind speed at a wind farm, which is then converted into the wind generation through the wind speed-power curve. In these studies, as the spatiotemporal correlation with other off-site locations is explicitly modeled, they exhibit a better performance opposed to those without considering it. However, as they primarily target at the applications of wind forecasting for a single wind farm, they cannot be applied for a large region where hundreds of wind farms are present.

This chapter presents an ensemble-based method for a TSO to produce a short-term wind power forecast, which combines the forecast output from a NWP model and the meteorological data from a network of weather stations. In contrast to the previous work in [21, 23], the proposed method takes advantage of the atmosphere models and capturing spatial and temporal correlation from multiple locations and thus is very suitable for producing wind forecast for a large footprint managed by a TSO. The contributions of this chapter are summarized as follows.

Its first novelty is to use meteorological observations at off-site weather stations over a large geographic region to complement the NWP model. For all of the preceding work [19–23], only the observation data at limited number of locations are used. Moreover, without a NWP, these methods also restrict themselves in capturing the spatial correlation in a region where no observation data can be found. In contrast, in this chapter, a network of over 100 weather stations in Texas, referred to as the West Texas Mesonet (WTM), will be used as the additional data to the forecast created by NWP. WTM has been originally installed for monitoring inclement weather conditions and will be continually expanded. As the weather physical process for a region like ERCOT is much more complex than a single WGR site, a combination of NWP and the WTM data can offer a great advantage in improving the accuracy of short-term wind forecasting for a TSO.

Compared to the previous work [19–23], a number of difficulties will arise when developing a model to blend the WTM data with the foresight of wind generation produced from NWP. Those difficulties are as follows: (1) the correlation between the WTM data and the wind power forecasting from NWP may vary dramatically over the time, (2) the model needs to handle a large amount of data from weather stations, (3) the model should be robust with a tolerance to some data errors or dropouts, and (4) the model should be adaptive to new scenarios which have not been seen in the past.

To address these difficulties, the second contribution of this chapter is to propose an ensemble-based method, which consists of three machine learning algorithms using the NWP output and WTM data as input. An ensemble method then obtains better prediction performance by strategically combining the results from those machine learning algorithms. Three machine learning algorithms (artificial neural network (ANN), SVR, GP) are proposed in this chapter to synthesize the data from NWP model and WTM as they have superior performance when dealing with non-stationary time-series data and an unknown model. By blending the results derived from three algorithms through a Bayesian model average (BMA), an ensemble forecast can be created with a performance superior to that of each individual algorithm. An ensemble method can also reduce the risk of over-estimation by preserving the diversity among all these models.

The proposed method is efficient and robust, and its performance has been validated and demonstrated by the 2-year operational data from ERCOT. The rest of this chapter is organized as follows. Section 2 discusses a region-level NWP model at ERCOT. The detailed description of WTM is presented in Sect. 3. Section 4 introduces the new method. The results based on the real-world ERCOT data are provided in Sect. 5, while the conclusions are given in Sect. 6.

2 Numerical Weather Prediction

ERCOT is an independent system operator serving over 23 million customers in Texas. As a single balancing authority without synchronous connections to its neighboring systems, ERCOT relies purely on its internal resources to balance power shortages and variations. The highest wind generation recorded at ERCOT was 16,141 MW on March 31, 2017. Since wind generation capacity is expected to be more than 25 GW by 2020, the role of WPF in reliable grid operations is becoming more critical at ERCOT. On the other hand, to forecast wind generation with precision is not a trivial task in Texas. The weather in the state of Texas can change rapidly so that wind generation is susceptible to severe changes caused by extreme weather. In addition, the installations of WGRs at ERCOT lack spatial diversity, i.e., a large number of WGRs are concentrated in the same geographic region.

A region-scale NWP is in operation at ERCOT to produce the hourly wind forecast every hour for the next 168 h [24]. It consists of four elements, namely, numerical simulation of atmosphere model, statistical models, plant output models, and ensemble optimization algorithm (EOA). Meteorological data at the sites of WGRs are used for setting the initial and lateral boundary conditions needed to run NWP models. The statistical models derive empirical predictive relationship from a historical data set that includes NWP model output, local generation, and meteorological measurement. They are also used to reduce NWP model biases. Plant output models are used to transform predictions of meteorological variables into predictions of wind generation. A forecast is generated from the combination of NWP models, statistical models, and plant output models. All of these forecasts are available as inputs to an EOA, which creates a composite forecast from the set of individual forecast by weighting them based on past performance and the expected conditions at the location of interest.

The accuracy and performance of the wind forecast are measured by its error. The 1-hour-ahead wind forecast error (WFE) for time t, $\delta_{t,\,t-1}$, is defined as the actual wind generation, y_t, subtracted by the 1-hour-ahead wind forecast, $f_{t,\,t-1}$, which is

$$\delta_{t,t-1} = y_t - f_{t,t-1} \tag{10.1}$$

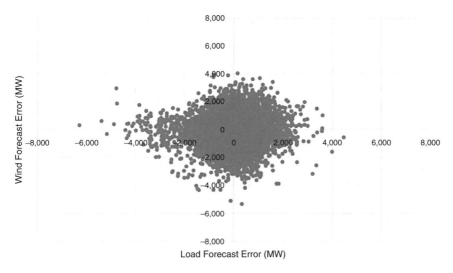

Fig. 10.1 3-hour-ahead load and wind forecast error in 2016

If divided by the wind installation capacity, the index to evaluate the forecasting performance is then converted into the mean absolute percent error (MAPE), which is

$$\varepsilon_{\text{eval}} = \sum_{t=1}^{Ne} \frac{|\delta_{t,t-1}|}{c_{\text{w},t} N_e} \tag{10.2}$$

Figure 10.1 depicts the scatter plot of 3-hour-ahead load forecast errors (LFE) and WFE assessed for each hour in 2016. In contrast to other cases where the penetration of WGRs is low, WFEs at ERCOT cannot be negligible as their magnitude is comparable to that of LFEs. For a majority of the time, WFEs fall into a range between −2000 MW and 2000 MW, indicating a satisfactory performance of the NWP model. However, there are a few instances when the large over-estimation error from NWP is greater than 5000 MW, thus resulting in a concern for the grid's reliability.

The 3-hour-ahead WFE time-series data is also non-stationary. It is asymmetric, is non-Gaussian, and has a fat-tail on both sides of its probability distribution function (PDF). The value of its kurtosis and skewness is 4.07 and −0.02, respectively. Thus, all of those need to be considered when designing a system to reduce or correct these errors.

3   West Texas Mesonet (WTM)

Besides the on-site meteorological data from WGRs, the data from the WTM is another source of information available to ERCOT. Figure 10.2 shows the locations of both WTM stations (blue) and WGRs (red). For economic reasons, the wind projects require a location with excellent wind resources so that the majority of WGRs at ERCOT are concentrated in a relatively small region of West Texas and the Texas Panhandle. In contrast, the locations of WTM stations cover a larger area as more than 100 stations have been installed. If the meteorological data can be collected at upstream WTM stations in the surrounding region of WGRs, it can help to detect large wind ramps earlier than only using the observations at down-stream WGRs. In this sense, the WTM data offers a better spatial and temporal correlation with WGRs.

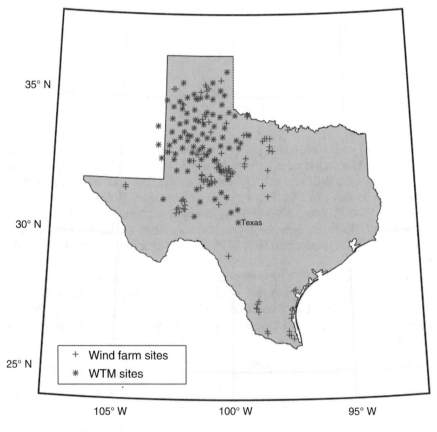

Fig. 10.2 Map of WTM locations

Each WTM station collects data every 1 to 5 minutes depending on the data logger and then sends the data collected to a central processing station. The data collected include:

1. 10-m wind speed and direction (average and 3-second peak wind speed)
2. 9-m temperature
3. 20-ft wind speed (fire weather) and 2-m wind speed
4. 2-m temperature
5. 1.5-m temperature and relative humidity (including dew point calculation)
6. Barometric pressure (using digital barometer: calculations include station pressure and altimeter)
7. Rainfall (total for the 5-minute period and an hourly summation product)
8. 2-m solar radiation

3.1   Value of WTM Data: An Example

As demonstrated in Sect. 2, large forecast errors from NWP models could occur due to defects in the models and incorrect boundary conditions. One of these examples is given in Fig. 10.3, where a large wind ramp-down event happened on May 29, 2015. Within 3 hours, the wind generation (in red) declined by 4 GW, and such a large magnitude of the change in wind generation was mainly driven by a large-scale

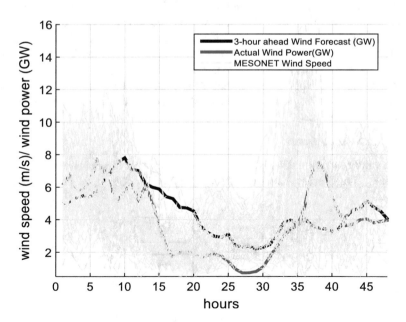

Fig. 10.3 3-hour-ahead wind forecast, actual wind generation, and wind speed from WTM on May 29 and May 30, 2015

weather phenomenon rather than a local turbulence. The 3-hour-ahead wind forecast produced by the NWP model predicted the ramp with a smaller magnitude. On the same day, the wind speed data from WTM (in yellow) already revealed a dramatic decrease in the wind speed (from hour 10 to 15) prior to the actual wind ramp-down (from hour 13 to hour 16). Similarly, a large wind ramp-up event was observed by WTM data (from hour 30 to hour 35) earlier before the actual event (from hour 35 to hour 40) occurred on the next day. This particular case demonstrated the capability of WTM data in observing a large wind ramp event earlier before it passed through WGRs. In other words, if 3-hour-ahead wind forecasting from the NWP model deviates largely from what WTM observes, these errors from NWP can be potentially reduced or corrected.

4   Machine Learning-Based Ensemble Method

As discussed previously, both NWP model and the WTM observation data are valuable in improving the accuracy of short-term wind forecasting. The former can handle the effect of complex terrains over the wind speed and local atmosphere turbulences and thus yields a satisfactory performance for the long prediction horizon. The WTM measurement is able to detect a dramatic change in the wind speed earlier before it propagates across WGRs. This chapter proposes a machine learning-based ensemble method which explores the temporal and spatial correlation between the wind forecast from NWP and the WTM observation data to produce a better wind forecasting. The historical wind forecast from NWP and the WTM observation data as well as the actual wind production are employed as time-series data to train and validate individual machine learning algorithm. The wind power forecast is ultimately calculated by BMA to combine the results derived from three heterogeneous predictors. The essential idea of the proposed method can be illustrated in Fig. 10.4. The proposed method mainly consists of three stages, which are detailed in what follows.

A. *Pre-processing Data*

The wind generation forecasting produced for the hour t from NWP at the time of $t - l$ (i.e., the forecast lead time is l), $x_{t,t-l}^{\mathrm{NWP}}$ ($x_{t,t-l}^{\mathrm{NWP}} \in \mathbb{R}^M$), is paired with the

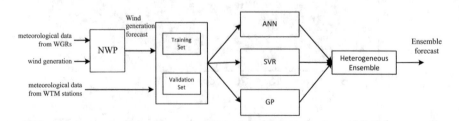

Fig. 10.4 Architecture and data flow of the proposed method

measurement data from WTM available at the time of t-l, x_{t-l}^{WTM} ($x_{t-l}^{WTM} \in \mathbb{R}^N$), and this pair is denoted as x_t. The actual wind generation produced at the time of t is y_t ($y_t \in \mathbb{R}$). Assume we have a training set TS of N_s observations, TS $= \{(x_t, y_t), t = 1,\ldots, N_s\}$, where x_t denotes an input vector of dimension D and y_t denotes the historical wind generation output for the time of t. As we have N_s samples, the whole input data, X, is represented by a $D \times N_s$ matrix, and with targets collected in a vector y, we can write the training data set as TS $= (X, y)$. Similarly, we have another test data set, VS, to validate the performance of each individual predictor and the proposed ensemble method.

B. *Machine Learning Algorithms*

In the second stage, three machine learning algorithms (predictors) are employed. The objective of each predictor, which is denoted as P_s ($s = 1,\ldots,S$), is to find out the relationship between x_t and y_t, which can be represented by

$$y_t = f_{P_s}(x_t) + \varepsilon_t \forall P_s \tag{10.3}$$

where f_{P_s} is a mapping function, $\mathbb{R}^D \to \mathbb{R}$, for the given predictor P_s and ε_t accounts for the unrepresented model or the process noise.

The base algorithms (predictors) should be chosen properly as they have a large influence on the overall performance of the ensemble method. First, each forecasting model should perform satisfactorily. Second, they should behave in an uncorrelated manner to preserve a diversity among them. There are a large number of forecasting methodologies and methods of time-series-based models available in the literature, which can be classified into three main categories: parametric, nonparametric, and machining learning [25]. As the wind generation is stochastic, the prior knowledge of the time series forecasted by NWP is not known, and its correlation with WTM data may be time varying. In this chapter, three machine learning algorithms (GP, SVR, and ANN) are chosen due to their capability in handling nonlinear and non-stationary data. The basic principles of those machine learning algorithms are described briefly in the following subsections.

1. *Gaussian Process* [17–18, 26]

A Gaussian process $f_{GP}(x)$ can be completely specified by its mean function and covariance function, written as $f_{GP}(x) \sim GP(m(x), k(x, x^T))$, where the mean function $m(x)$ and covariance function $k(x, x^T)$ are defined as

$$m(x) = E[f_{GP}(x)] \tag{10.4}$$

$$k(x, x^T) = E[(f_{GP}(x) - m(x))(f_{GP}(x^T) - m(x^T))] \tag{10.5}$$

When a Gaussian process is used to model the relationship between input x_t and target y_t, the function satisfies

$$y_t = f_{\text{GP}}(x_t) + \tau_t \tag{10.6}$$

where additive noise τ_t is assumed to be an independent, identically distributed Gaussian distribution with zero mean and variance σ_n^2, i.e., $\tau_t \sim N(0, \sigma_n^2)$. Note that y is a linear combination of Gaussian variables and hence is itself Gaussian. The prior on y becomes

$$E[y] = E[f_{\text{GP}}(X)] = E_y \tag{10.7}$$

$$\text{cov}[y] = K(X, X) + \sigma_n^2 I \tag{10.8}$$

where K is a matrix with elements $K_{ij} = k(x_i, x_j)$, which is also known as the kernel function.

Given a training set TS $= (X, y)$, the goal is to make prediction of the target variable f_{GP}^* for a new input x^*. Since we already have

$$p(y|X, k) = N(E_y, K + \sigma_n^2 I) \tag{10.9}$$

the distribution with new input can be written as

$$\begin{bmatrix} y \\ f_{\text{GP}}^* \end{bmatrix} \sim N\left(E_y, \begin{bmatrix} K(X,X)+\sigma_n^2 I & k(X,x^*) \\ k(x^*,X) & k(x^*,x^*) \end{bmatrix}\right) \tag{10.10}$$

where $k_* = k(X, x^*) = k(x^*, X)^T = [k(x_1, x^*), \ldots, k(x_{N_s}, x^*)]$. Then according to the principle of joint Gaussian distributions, the prediction for the target is given by

$$f_{\text{GP}}^* = k_*^T (K + \sigma_n^2 I)^{-1} y \tag{10.11}$$

$$\text{cov}[f_{\text{GP}}^*] = k(x^*, x^*) - k_*^T (K + \sigma_n^2 I)^{-1} k_* \tag{10.12}$$

2. SVR [27, 28]

SVR is considered a nonparametric technique because it relies on kernel functions and has been widely used for regression problems. Given a training set TS $= (X, y)$, we have a set of data where X is a multivariate set of N_s observations, x_t, with observed response values y_t. The problem is formulated as to find the linear function

$$f_{\text{SVR}}(x_t) = x_t^T \beta + b \tag{10.13}$$

and ensure that is as flat as possible, i.e., to find $f_{\text{SVR}}(x_t)$ with the minimum norm value. This is solved as a convex optimization problem to minimize

$$J(\beta) = \frac{1}{2}\beta^T\beta \tag{10.14}$$

subject to all residuals having a value less than ϵ; or, in equation form

$$\left|y_t - \left(x_t^T\beta + b\right)\right| \le \epsilon \forall t \tag{10.15}$$

Once the training is completed, for a new input, x^*, the point prediction produced by SVR is

$$f^*_{SVR} = (x^*)^T\beta + b \tag{10.16}$$

3. *ANN* [29]

An ANN learns from given sample examples, by constructing an input-output mapping to perform predictions of future samples. ANN basically consists of three layers: an input layer, a hidden layer, and an output layer. Suppose x_t denotes an input vector of dimension D, $\omega_{m,n}$ are the weightings between the input layer and the hidden layer, and v_n are the weightings between the hidden layer and the output layer. y_t is the expected output. The output of the hidden layer with an input of x_t can be formulated as

$$z_{n,t} = g\left(\sum_{m=1}^{D}\omega_{m,n}x_t^m\right) \tag{10.17}$$

where g is the transfer function to handle the nonlinear problem.

Similarly, the output of the output layer is

$$u_t = \sum v_n z_{n,t} \tag{10.18}$$

Then the error signal e_t is produced as

$$e_t = \frac{1}{2}(y_t - u_t)^2 \tag{10.19}$$

The error function, e_t, measures the discrepancy between the desired output, y_t, and the output computed by the ANN, u_t, at time t. When the ANN is trained in the supervised manner, the weights $\omega_{m,n}$ and v_n are adjusted to minimize this error function (10.19).

C. *Ensemble Forecast by BMA*

In the third stage, a BMA method is used to build an ensemble forecasting model. An ensemble prediction is given by combining the predictions of S predictors f_{P_s} with $s \in \{1,...,S\}$. A comprehensive overview and empirical analysis for the ensemble method is given by Bauer and Kohavi [30]. In particular, BMA is proposed in this chapter as a way of combining forecasts from different sources into a consensus PDF [31–33]. The combined forecast PDF of wind generation, y, given the data, X, is

$$p(y|X) = \sum_{s=1}^{S} p(y|P_s, X) p(P_s|X) \qquad (10.20)$$

where $p(y| P_s, X)$ is the forecast PDF based on model P_s alone, estimated from the training data, and $p(P_s| X)$ is the posterior probability of model P_s being correct given the training data, X. The term is computed with the aid of Bayes's theorem

$$p(P_s|X) = \frac{p(X|P_s)p(P_s)}{\sum\limits_{s=1}^{S} p(X|P_s)p(P_s)} \qquad (10.21)$$

$p(P_s|X)$ is the BMA weight for the model s, computed from the training data set, and reflects the relative performance of model s on the training period. The weights, $p(P_s|X)$, add up to 1. Successful implementation of the BMA method requires estimates of these weights, and their values can be estimated by maximizing a likelihood function from the training data. As there is no closed-form solution to maximize the likelihood function, an iterative solution method such as the expectation-maximization algorithm can be used. Mathematically, once the weights are determined, the ensemble wind forecast is given by

$$y_t^{en} = \sum_{s=1}^{S} (P_s|X) f_{P_s}(x_t), \quad t = 1, \ldots, Ns \qquad (10.22)$$

5   Experimental Results

By the end of the year 2017, about 21,000 MW of WGRs have been installed in the ERCOT region. As WGRs account for a large portion of generation portfolio at ERCOT, an accurate wind forecasting is critical to both market operation and reliability management of the grid. In 2017, the MAPE of day-ahead and hour-ahead wind forecasting from NWP model at ERCOT is 5.56% and 3.6%, respectively. As a more accurate wind forecasting can bring tremendous benefits to the grid operation, it is estimated that an improvement over the forecasting accuracy by 10% could save ERCOT millions of dollars annually.

The performance of the proposed model was tested and validated using 2-year ERCOT operational data to demonstrate its effectiveness. The ERCOT operational data is divided into a 1-year training data set, TS, and a 1-year test data set, VS. The wind forecasting from NWP in 2015 and the WTM data in the same year are used as the training data set, TS. Once the training is completed, the same data set in 2016 is applied as the test data, VS, to evaluate the accuracy of wind forecasting for the proposed method. The forecasting data produced by NWP was collected from a third-party wind forecasting service provider at ERCOT, which is the rolling wind forecast for the next 168 hours updated each hour. To account for the difference in wind installation capacity over the time, both the actual wind production and the wind forecasting are normalized with respect to the corresponding wind installation capacity.

The performance of the proposed method also depends on how to choose the appropriate set of the input variables. While the WTM data sets include eight different types of data, only those variables which exhibit a strong degree of statistical correlation with wind generation should be included. Thus, only wind speed and wind direction are chosen from the WTM data as part of the input to the ensemble-based model. Currently, time-of-day and seasonal variations in the atmospheric boundary layers and turbulence are not considered, but will be addressed in the future work.

Depending on the time scale of the applications, the wind forecasting horizon can range from minutes to weeks. In the past, significant efforts have been made to increase the accuracy of the short-term WPF with a forecasting horizon of from 1- to 48-hour ahead. From the perspective of grid operations, 3-hour-ahead forecasting horizon is the most important since the wind forecasting can be adjusted based on the meteorological information close to the operation hour, while operators still have time to respond to the changes foreseen in the weather [9, 34]. Moreover, the amount of non-spinning reserve service needed at ERCOT is sized based on 3-hour-ahead load and wind forecasting errors [9]. In this chapter, the effectiveness of the proposed method is demonstrated for 3-hour-ahead wind forecasting although it can be generally applied to other forecasting horizons.

Figure 10.5 depicts the cross-correlation between wind speed measured at different weather stations and the ERCOT-aggregated wind generation. It shows a strong dependence between two variables, with a cross-correlation between 0.32 and 0.92. Also, due to a geographic diversity of WTM, the time lags between those two (when their cross-correlation is at peak) can vary from 12 hours (lagging) to 8 hours (leading).

5.1 Performance of Three Machine Learning Algorithms

After the base algorithms have been trained, the MAPE of 3-hour-ahead wind forecasting for the ANN-, SVR-, and GP-based method using 2016 test data set is 4.06%, 3.87%, and 3.81%, respectively, compared to 4.57% if only NWP is in use, as shown in Table 10.1. All of three base algorithms outperform the NWP delivered by the third-party wind forecasting service vendor at ERCOT, and the corresponding improvement on accuracy over the NWP ranges from 11.20% to 16.60%. The

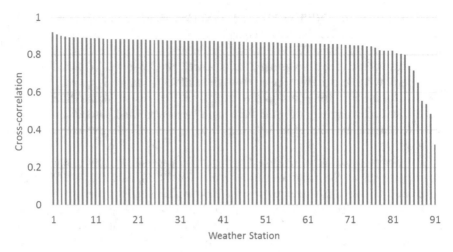

Fig. 10.5 Cross-correlation between wind speed at WTM stations and ERCOT wind generation

Table 10.1 MAPE of 3-hour-ahead wind forecast using 2016 test data set

Method	MAPE	Improvement over NWP model
NWP only	4.57%	–
ANN	4.06%	11.20%
SVR	3.87%	15.30%
GP	3.81%	16.60%
Ensemble forecast	3.74%	18.20%

improvement of 3-hour-ahead wind forecasting accuracy confirms that (1) three machine learning algorithms can process non-stationary data and (2) the WTM data can reveal the temporal and spatial correlation between the weather stations and wind farms. Among the three machine learning algorithms, GP-based method yields the best result, while the performance of ANN- and SVR-based methods is also satisfactory. The forecasting from these base algorithms can be further blended to build an ensemble forecast, which will be discussed next.

The PDFs of the forecasting errors for the ANN-, SVR-, and GP-based model as well as NWP are depicted in Fig. 10.6. The negative range represents the over-forecasting of wind generation, while the positive range denotes the under-forecasting of wind generation. All of the PDFs in Fig. 10.6 have a similar shape with heavy tails. One notable feature is that the machine learning algorithms have shifted the forecasting errors slightly to the right side (under-forecasting) compared to the NWP model. Among the three machine learning approaches, the PDFs of their forecasting errors are almost indistinguishable, which is consistent with their similar forecasting performance in terms of MAPE.

Fig. 10.6 PDF of 3-hour-ahead wind forecast errors in 2016

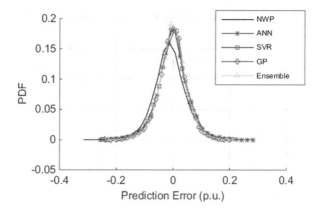

Table 10.2 MAPE of 3-hour-ahead wind forecast for large wind ramp events

	Large wind ramp-down event	Improvement over NWP model
NWP only	6.09%	–
ANN	5.87%	3.60%
SVR	5.75%	5.60%
GP	5.71%	6.20%
Ensemble forecast	5.58%	8.40%

5.2   Performance of Base Algorithms for Large Wind Ramp

The performance of the proposed method has been also evaluated for large wind ramp-down events, which are defined as a drop in the total wind generation greater than 4000 MW within 3 hours. Large wind ramp events are susceptible to large forecasting errors so that the MAPE of the NWP model is increased to 6.09% compared to its performance evaluated for all of the hours (4.57%).

For those events, the machine learning algorithms perform better than the NWP model, and the improvement ranges from 3.6% to 8.4% as shown in Table 10.2. As the proposed methods lead to less over-forecasting errors for large wind ramp events, this partially alleviates the reliability concern resulting from the over-forecasting of wind generation. If only the over-forecasting periods for large wind ramp-down events are considered, the performance of the ANN-, SVR-, and GP-based method is 25.3%, 24.5%, and 23.8% better than that of the NWP model, respectively, as shown in Table 10.3.

5.3   Performance of Ensemble Method

The proposed ensemble method is more accurate than the individual machine learning algorithm (see Tables 10.1, 10.2, and 10.3), and its detection of large

Table 10.3 MAPE of 3-hour-ahead wind forecast for large wind ramp events when over-forecasting

	Large wind ramp-down event and over-forecasting	Improvement over NWP model
NWP only	6.70%	–
ANN	5.00%	25.30%
SVR	5.00%	24.50%
GP	5.10%	23.80%
Ensemble forecast	4.80%	28.60%

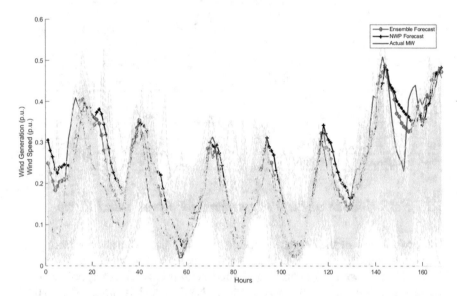

Fig. 10.7 3-hour -ahead wind generation forecast, actual wind production, and wind speed at WTM (yellow dot lines) in p.u. for 7 days in 2016 when evaluated for forecasting performance

wind ramp events is also more effective. One typical 7-day profile of 3-hour-ahead wind forecast and wind speed at WTM in 2016 is given in Fig. 10.7. The wind speed at WTM is normalized with a base of 20 m/s. For this particular period of the time, the proposed ensemble forecast is able to better predict the actual fluctuations of wind generation so that the forecast errors are smaller compared to the forecast created by NWP.

5.4   Robustness of Proposed Method

The robustness of the ensemble-based method was tested by removing the data at some selected WTM weather stations from the input. It was found that the proposed model is able to tolerate the simultaneous dropout of the data from any of two weather stations without significantly impacting the forecasting performance as the resulting change in MAPE is less than 0.04%.

6 Conclusions

The wind forecasting can bring tremendous benefits to the reliability and efficiency of a grid especially when the penetration of wind generation resources is high. An accurate wind forecasting is widely considered as one of the prerequisites to the reliable integration of a large amount of WGRs into the power grid. A NWP model is used at ERCOT to produce the wind forecasting for the next 168 hours. The further improvement on the accuracy of the wind forecasting could help ERCOT to save millions of dollars each year.

An ensemble method has been proposed in this chapter to improve the performance of short-term wind forecasting. It uses three machine learning algorithms to take advantage of data from a network of weather stations to reduce the errors of the wind forecasting produced from a NWP model. The proposed method is also more effective and robust as it uses a BMA to combine the results from three machine learning algorithms. Since the network of weather stations can cover a larger region than WGRs, they can provide an earlier detection of large wind ramp before the changes in the wind speed propagate the locations of WGRs. Thus, the proposed approach is advantageous by combining the strength of the simulation-based NWP model and the measurement-based WTM data.

The proposed method has been trained by the 1-year historical data and then validated by another 1-year test data collected at ERCOT. The experimental results show that the proposed ensemble method can improve the performance of 3-hour-ahead wind forecasting from 4.57% to 3.74% in terms of MAPE compared to NWP model. In addition, the test results also show that an ensemble forecast is superior to any of three machine learning algorithms. The gain in the performance improvement is even larger if only over-forecasting periods for large wind ramp-down events are considered.

References

1. Pengwei Du, Ross Baldick, and Aidan Tuohy, Integration of Large Scale Renewable Energy into Bulk Power Systems: From Planning to Operation, Springer, 2017.
2. DOE, 20% Wind Energy by 2030 (Washington, DC: DOE, 2008).
3. C. Böhringer, A. Löschel, U. Moslener, and T.F. Rutherford, "EU climate policy up to 2020: An economic impact assessment," *Energy Economics*, vol. 31, pp. 295–305, 2009.
4. Pengwei Du, Hailong Hui, and Ning Lu, "Procurement of regulation services for a grid with high-penetration wind generation resources: a case study of ERCOT," *IET Generation, Transmission & Distribution*, vol. 10, pp. 4085–4093, 2016.

5. González-Aparicio, I., and A. Zucker, "Impact of wind power uncertainty forecasting on the market integration of wind energy in Spain," *Applied Energy*, vol. 159, pp. 334–349, 2015.
6. Yuri V Makarov, Pengwei Du, "Sizing energy storage to accommodate high penetration of variable energy resources," *IEEE Transactions on sustainable Energy*, vol. 3, pp. 34–40, 2012.
7. Pengwei Du, and Ning Lu, Energy Storage for Smart Grids: Planning and Operation for Renewable and Variable Energy Resources (VERs), Academic Press, 2014.
8. C. Monteiro, R. Bessa, V. Miranda, A. Botterud, J. Wang, and G. Conzelmann, Wind power forecasting: state-of-the-art 2009, No. ANL/DIS-10-1, Argonne National Laboratory (ANL), 2009.
9. Nischal Rajbhandari, Weifeng Li, Pengwei Du, Sandip Sharma, and Bill Blevins, "Analysis of net-load forecast error and new methodology to determine Non-Spin Reserve Service requirement," IEEE Power and Energy Society General Meeting (PESGM), 2016, 1–5.
10. Wang, J., A. Botterud, R. Bessa, H. Keko, L. Carvalho, D. Issicaba, J. Sumaili, and V. Miranda, "Wind power forecasting uncertainty and unit commitment," *Applied Energy*, vol. 88, pp. 4014–4023, 2011.
11. Al-Yahyai, Sultan, Yassine Charabi, and Adel Gastli, "Review of the use of numerical weather prediction (NWP) models for wind energy assessment," *Renewable and Sustainable Energy Reviews*, vol. 9, pp. 3192–3198, 2010.
12. Möhrlen, Corinna, and Jess U. Jørgensen, The Role of Ensemble Forecasting in Integrating Renewables into Power Systems: From Theory to Real-Time Applications, In Integration of Large-Scale Renewable Energy into Bulk Power Systems, Springer, Cham, 2017.
13. Cassola, Federico, and Massimiliano Burlando, "Wind speed and wind energy forecast through Kalman filtering of Numerical Weather Prediction model output," *Applied energy*, vol. 99, pp. 154–166, 2012.
14. Shengchen Fang, and Hsiao-Dong Chiang, "A high-accuracy wind power forecasting model," *IEEE Transactions on Power Systems*, vol. 32, pp. 1589–1590, 2017.
15. Andrade José R., and Ricardo J. Bessa, "Improving renewable energy forecasting with a grid of numerical weather predictions," *IEEE Transactions on Sustainable Energy*, vol. 4, pp. 1571–1580, 2017.
16. Zhao Yongning, Lin Ye, Pierre Pinson, Yong Tang, and Peng Lu. "Correlation-constrained and sparsity-controlled vector autoregressive model for spatio-temporal wind power forecasting," *IEEE Transactions on Power Systems*, 2018.
17. Niya Chen, Zheng Qian, Ian T. Nabney, and Xiaofeng Meng, "Wind power forecasts using Gaussian processes and numerical weather prediction," *IEEE Transactions on Power Systems*, vol. 29, pp. 656–665, 2014.
18. Duehee Lee, and Ross Baldick, "Short-term wind power ensemble prediction based on Gaussian processes and neural networks," *IEEE Transactions on Smart Grid*, vol. 5, pp. 501–510, 2014.
19. Miao He, Lei Yang, Junshan Zhang, and Vijay Vittal, "A spatio-temporal analysis approach for short-term forecast of wind farm generation," *IEEE Transactions on Power Systems*, vol. 4, pp. 1611–1622, 2014.
20. Yao Zhang, Jianxue Wang, "A distributed approach for wind power probabilistic forecasting considering spatio-temporal correlation without direct access to off-site information," *IEEE Transactions on Power Systems*, 2018.
21. Shu Fan, James R. Liao, Ryuichi Yokoyama, Luonan Chen, and Wei-Jen Lee, "Forecasting the wind generation using a two-stage network based on meteorological information," *IEEE Transactions on Energy Conversion*, vol. 24, pp. 474–482, 2009.
22. Ezzat Ahmed Aziz, Mikyoung Jun, and Yu Ding, "Spatio-temporal asymmetry of local wind fields and its impact on short-term wind forecasting," *IEEE Transactions on Sustainable Energy*, 2018.
23. M.C Alexiadis, P.S. Dokopoulos, and H.S. Sahsamanoglou, "Wind speed and power forecasting based on spatial correlation models," *IEEE Transactions on Energy Conversion*, vol. 14, pp. 836–842, 1999.

24. AWS Truepower user manual, 2009.
25. G. E. P. Box, G. M. Jenkins, and G. C. Reinsel, Time Series Analysis: Forecasting and Control, Prentice Hall, Englewood Cliffs, NJ, 1994.
26. C. E. Rasmussen, and C. K. I. Williams, Gaussian Processes for Machine Learning, MIT Press, Cambridge, Massachusetts, 2006.
27. V. Vapnik, The Nature of Statistical Learning Theory, Springer, New York, 1995.
28. C.-C. Chang and C.-J. Lin. LIBSVM, "A library for support vector machines," *ACM Transactions on Intelligent Systems and Technology*, 1–27, 2011.
29. Li Gong, and Jing Shi, "On comparing three artificial neural networks for wind speed forecasting," *Applied Energy*, vol. 7, pp. 2313–2320, 2010.
30. E. Bauer and R. Kohavi, "An empirical comparison of voting classification algorithms: Bagging, boosting, and variants," *Machine learning*, vol. 36, pp. 105–139, 1999.
31. J.A. Vrugta, MODELAVG, "A MATLAB Toolbox for Postprocessing of Model Ensembles," 2016, http://faculty.sites.uci.edu/jasper/files/2016/04/manual_Model_averaging.pdf.
32. Casanova Sophie, and Bodo Ahrens, "On the weighting of multimodel ensembles in seasonal and short-range weather forecasting," *Monthly weather review*, vol. 11, pp. 3811–3822, 2009.
33. Wilson Laurence J., Stephane Beauregard, Adrian E. Raftery, and Richard Verret, "Calibrated surface temperature forecasts from the Canadian ensemble prediction system using Bayesian model averaging," *Monthly Weather Review*, vol. 4, pp. 1364–1385, 2007.
34. ERCOT Nodal Operating Guide, http://www.ercot.com/mktrules/guides/noperating, 2018.

Index

A
Ancillary services, 13, 27, 76, 103, 140, 209

C
Critical inertia, 23, 192–193, 195, 205–209,
    212–222

D
Day-ahead market (DAM), 13–15, 18, 22,
    30–48, 50, 52, 55, 73, 77–86,
    100, 109, 115, 125, 140–142,
    153, 154, 158, 179

E
Electric Reliability Council of Texas (ERCOT),
    1–25, 27–44, 47, 48, 51–53, 55, 61,
    63–67, 71–73, 75–78, 81, 83–89, 91, 92,
    94, 95, 100, 101, 103–113, 116, 118,
    120–126, 130, 131, 133, 134, 148, 155,
    156, 162–171, 179–185, 187–190, 192,
    193, 195, 199–205, 208–218, 220–222,
    224, 225, 227, 228, 233, 235, 238, 241,
    243–250, 254, 255, 262, 265–268,
    274–276, 279
Ensemble forecast, 251, 254, 265, 273,
    276–279
ERCOT Market, 1, 14, 27, 37, 73, 81, 205

F
Fast frequency response (FFR), 16, 118–123,
    140–142, 144, 146, 148–154, 171,
    177–195, 209, 213–215, 219–222
5-minute wind and solar ramp, 101
Frequency response reserve (FRR)
    market, 171

G
Generators, 2, 7, 9, 11, 15, 17–19, 23, 24, 66,
    71, 103–105, 109, 110, 112, 118–121,
    137–172, 178–184, 190, 193, 194,
    199–201, 219, 221, 230

I
Inertia-less wind and solar resources, 134

L
Load resources with under-frequency relays,
    118, 222

M
Meteorological observations, 265
Mixed-integer linear programming
    (MILP), 56, 156
Multiple-Period Reactive Power Coordination,
    224–241

© The Editor(s) (if applicable) and The Author(s), under exclusive license to   283
Springer Nature Switzerland AG 2023
P. Du, *Renewable Energy Integration for Bulk Power Systems*, Power Electronics
and Power Systems, https://doi.org/10.1007/978-3-031-28639-1

N

New Ancillary Service Market, 177–195
Non-spinning reserves, 14, 15, 17, 23, 77,
        86, 103, 118, 133, 134, 141, 142,
        144, 145, 148, 275
Non-synchronous resources, 19, 199
Non-to-exceedance (NTE), 89–94, 101
Numerical weather prediction (NWP), 243,
        250, 251, 253–256, 258, 263–279

O

Operational security, 7, 24

P

Primary frequency control (PFC), 118,
        137–172, 179, 181–184, 195

R

Real-time market, 15, 30, 32, 34, 35, 37
Real-time SCED, 73, 78, 81–84, 106
Regulation down, 14, 16, 72, 77, 133, 144, 148
Regulation up, 14, 16, 46, 47, 72, 77, 94, 111,
        118, 133, 142, 144, 145, 148
Reliability unit commitment (RUC), 13, 14, 22,
        30, 36, 39, 42, 47–63, 73, 77–86, 100,
        106, 110, 211, 221, 230, 244, 255
Renewable energy forecasting, 243
Renewable integration, 73, 178, 227–228

Renewable resources, 5, 12, 17, 20, 73, 75–100,
        125, 131, 139–140, 157, 171, 172, 193,
        225, 226, 243, 244
Responsive reserves, 14, 15, 18, 44–46, 70,
        72, 77, 103, 118–123, 133, 134, 145,
        180, 181

S

System inertia, 12, 16, 19, 23, 103, 112,
        118–123, 139, 146, 148, 150–152,
        156–158, 160, 161, 163, 167, 171,
        178, 179, 181–184, 186, 190, 191,
        193–195, 199–222

T

Transmission Development and Capacity
        Adequacy, 9–13

U

Under-frequency Load Shedding (UFLS), 23,
        118–120, 181, 184, 190, 192, 193,
        208–210, 212, 214–217, 219, 222

W

Wholesale electricity market, 1, 7, 75, 177
Wind generation forecasting, 106–108,
        264, 270